Praise for

EXTRAORDINARY INSECTS

"Anne Sverdrup-Thygeson writes about insects with such enthusiasm and affection, you'll wish you were an entomologist! But it's never too late to develop a love for bugs, and *Extraordinary Insects* is the perfect guide—filled with surprising, fascinating, and often extremely funny stories."

—Thor Hanson, author of
Buzz: The Nature and Necessity of Bees

"We live on the planet of insects, and Sverdrup-Thygeson brings it to life in this sharp, good-humored presentation. . . . A classy and brightly informative appreciation of insects—all you could ask for in a popular natural history."

—*Kirkus Reviews*

"Insects—from jewel beetles to stink bugs—present an embarrassment of riches, as ecologist Anne Sverdrup-Thygeson spectacularly proves. She tours their anatomy, reproduction and more, delivering a hail of facts with brio and precision."

—Barbara Kiser, Nature.com

"Sverdrup-Thygeson is a lively, witty, and discerning guide through the scientific lore surrounding some of the tiniest—though still very powerful—organisms on Earth."

—Michael Berry, *Sierra Club Magazine*

"A cornucopia of fascinating insights that paint a portrait of the deep interconnectedness of human culture with our six-legged friends."

—Lynn Dicks, *Science*

"Most people can easily list all the ways insects annoy them, but it turns out that our lives are made immeasurably better by insects. Sverdrup-Thygeson does a great job outlining the disproportionately large impact insects have on people's daily lives, in proportion to their small size."

—Robin Thomson, *Minneapolis Star-Tribune*

"A very enthusiastic look at the flying, crawling, stinging bug universe world, and why we should cherish it."

—John Timpane, *The Philadelphia Inquirer*

"[*Extraordinary Insects*] avoids the slippery slope of boring, teachy prose by leaning into the weirdness of the bug world. Sverdrup-Thygeson writes like a scientist but in a spare, charming way."

—Heather Hansman, *Outside* (online)

"Anne Sverdrup-Thygeson writes playfully and accessibly about the world of insects. . . . This book is fascinating, so fun."

—Arianna Rebolini, *Buzzfeed*

"Conservation biologist Sverdrup-Thygeson exudes an infectious enthusiasm for all things entomological in this curiosity-provoking primer. . . . Playful and evocative."

—*Publishers Weekly* (starred review)

"A lively introduction to the six-legged creatures that share our planet. . . . *Extraordinary Insects* will foster affection for its winged, creeping, and crawling subjects, even among its most bug-shy readers."

—*Booklist*

EXTRAORDINARY INSECTS

The Fabulous, Indispensable Creatures Who Run Our World

Previously published as
Buzz, Sting, Bite: Why We Need Insects

Anne Sverdrup-Thygeson

Translated by Lucy Moffatt
Illustrations by Tuva Sverdrup-Thygeson

Simon & Schuster Paperbacks
NEW YORK · LONDON · TORONTO
SYDNEY · NEW DELHI

Simon & Schuster Paperbacks
An Imprint of Simon & Schuster, Inc.
1230 Avenue of the Americas
New York, NY 10020

Originally published in 2018 in Norway by J.M. Stenersens Forlag
as *Insektenes Planet*

This translation has been published with the financial support of NORLA,
Norwegian Literature Abroad

Previously published as *Buzz, Sting, Bite*

First Simon & Schuster paperback edition July 2020

Interior design by Paul Dippolito

Manufactured in the United States of America

3 5 7 9 10 8 6 4

The Library of Congress has cataloged the hardcover edition as follows:

Names: Sverdrup-Thygeson, Anne, author. | Moffatt, Lucy, translator.
Title: Buzz, sting, bite : why we need insects / by Anne Sverdrup-Thygeson ;
translated by Lucy Moffatt.
Other titles: Insektenes planet. English
Description: First Simon & Schuster hardcover edition July 2019. | New York :
Simon & Schuster, 2019. | "Originally published in 2018 in Norway by J.M.
Stenersens Forlag as Insektenes Planet." | Includes bibliographical references and index.
Identifiers: LCCN 2018059668| ISBN 9781982112875 | ISBN 1982112875
Subjects: LCSH: Insects.
Classification: LCC QL463 .S82 2019 | DDC 595.7--dc23 LC record available at
https://lccn.loc.gov/2018059668.

ISBN 978-1-9821-1287-5
ISBN 978-1-9821-1288-2 (pbk)
ISBN 978-1-9821-1289-9 (ebook)

To Karine, Tuva and Simen—I hope your generation takes better care of our six-legged lifesavers. . .

Nature is nowhere as great as in
its smallest creatures.

PLINY THE ELDER,
HISTORIA NATURALIS II, I.4,
C. 79 CE

Contents

Preface

I've always liked spending time outdoors, especially in the forest—preferably in places where signs of human life are few and far between and evidence of our modern impact scarce; among trees older than any living person, trees that have toppled headlong, nose-diving into the springy moss. Here they lie, in prostrate silence, as life continues its eternal round dance.

The insects come to the dead tree in hordes. Bark beetles party in the sap that ferments beneath the bark, longhorn beetle larvae trace ingenious patterns on the surface of the wood, and wireworms like tiny crocodiles greedily gobble anything that moves within the rotting wood. Together, thousands of insects, fungi, and bacteria work to break down dead matter and transform it into new life.

I feel incredibly lucky to be able to research such an exciting topic. I have a fantastic job: I am a professor at Norway's University of Life Sciences, where I work as a scientist, teacher, and communicator. One day I might be reading about new research, digging deep and losing myself in scientific detail. The next, I'm due to give a lecture and have to look for an overarching structure in a given subject area—find examples and illustrate why the issue matters to you and me. Maybe it will end up as a post on our research blog, *Insektøkologene* (*The Insect Ecologists*).

Sometimes I work outdoors. I seek out ancient, hollow oaks or map forests affected to various degrees by logging. All this I do in the company of my wonderful colleagues and students.

Preface

When I tell people I work with insects, they often ask me: What good are wasps? Or: Why do we need mosquitoes and deer flies? Some insects are a nuisance, of course. But they are a vanishingly small minority compared with the myriads of tiny critters that do their little bit to save your life. But I have three answers for the troublesome ones.

First, these annoying insects are useful to nature. Mosquitoes, gnats, and their relatives are vital food for fish, birds, bats, and other creatures. In the highlands and the far north, in particular, swarms of flies and mosquitoes are crucial to animals much larger than themselves, and on a vast scale. During the short, hectic Arctic summer, insect swarms can determine where the large reindeer flocks graze, trample the earth, and deposit nutrition in the form of dung. This has ripple effects that influence the whole ecosystem. Similarly, yellow jackets are useful, both for us and other creatures. They help pollinate plants, gobble up pests whose numbers we'd rather keep down, and provide food for honey buzzards and countless other species.

Second, helpful solutions may await us where least we expect them. This applies even to creatures we see as yucky nuisances. Blowflies can cleanse hard-to-heal wounds, while mealworms turn out to be able to digest plastic, and scientists are currently investigating the use of cockroaches for rescue work in collapsed or severely polluted buildings.

Third, many people think that all species should have the opportunity to achieve their full life potential—that we humans have no right to play fast and loose with species diversity driven by shortsighted judgments about which species we see as cute or useful. This means we have a moral duty to take the best possible care of our planet's myriad creatures—including those that do not engage in any visible value creation, insects that do not have soft fur and big brown eyes, and species we see no point in.

Nature is bewildering in its complexity, and insects are a significant part of the ingeniously constructed systems in which we humans are just one species among millions of others. That is why this book will deal with the very smallest among us: all the strange, beautiful, and bizarre insects that underpin the world as we know it.

The first part of the book is about the insects themselves. In the first chapter, you can read about their mind-bogglingly rich variety, how they are put together, how they sense their surroundings, and a bit about how to recognize the most important insect groups. Then in chapter 2, you'll gain an insight into their rather strange sex lives.

After that, I'll dig deeper into the intricate interplay between insects and other animals (chapter 3) and between insects and plants (chapter 4): the daily struggle to eat or be eaten, in which every creature battles to pass on its own genes. Yet there is still room for collaboration, in countless peculiar ways.

The rest of the book is about insects' intimate relationship with one particular species: us—how they contribute to our food supply (chapter 5), clean up the natural environment (chapter 6), and give us things we need, from honey to antibiotics (chapter 7). In chapter 8, I take a look at new fields in which insects can lead the way. Finally, in chapter 9, I consider how these tiny helpers of ours are getting along and how you and I can help improve their lot. Because we humans rely on insects getting their job done. We need them for pollination, decomposition, and soil formation; to serve as food for other animals, keep harmful organisms in check, disperse seeds, help us in our research, and inspire us with their smart solutions. Insects are nature's little cogs that make the world go round.

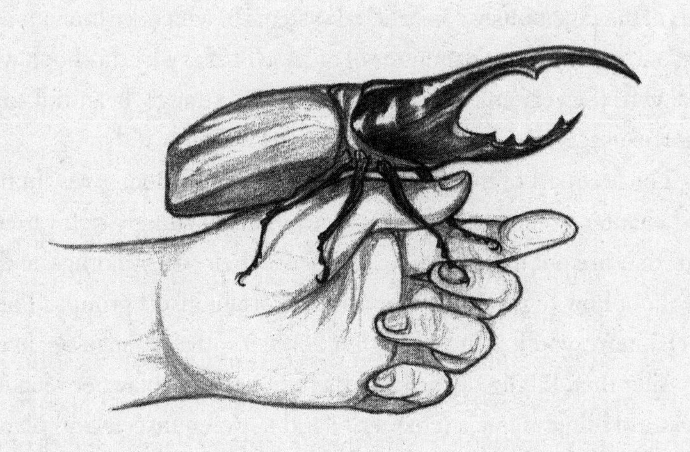

Introduction

There are more than 200 million insects for every human being living on the planet today. As you sit reading this sentence, between 1 quintillion and 10 quintillion insects are shuffling and crawling and flapping around on the planet, outnumbering the grains of sand on all the world's beaches. Like it or not, they have you surrounded, because Earth is the planet of the insects.

There are so many of them that it's difficult to take it in, and they are everywhere: in forests and lakes, meadows and rivers, tundra and mountains. Stone flies live in the chilly heights of the Himalayas at altitudes of 20,000 feet, while brine flies live in the piping hot springs of Yellowstone National Park, where temperatures exceed 120 degrees Fahrenheit. In the eternal darkness of the world's deepest caverns live blind cave midges. Insects can live in baptismal fonts, computers, oil puddles, and the acid and bile of a horse's stomach. They live in deserts, beneath the ice on frozen seas, in the snow, and in the nostrils of walruses.

Insects live on all continents—although they are admittedly represented by only a single species on Antarctica: a flightless midge that can't survive if the temperature happens to creep up to 50 degrees for any length of time. There are even insects in the sea. Seals and penguins have in their hides various kinds of lice, which remain in place when their hosts dive beneath the surface. And we mustn't forget the lice that live in a pelican's pouch or the water striders who spend their lives scudding six-legged across the open sea.

Insects may be tiny, but their achievements are far from trifling. Long before human beings set foot on this planet, insects had already taken up agriculture and animal husbandry: termites grow fungus for food, while ants keep aphids as dairy cattle. Wasps were the first creatures to make paper from cellulose, and caddis fly larvae were catching other creatures in netlike webs millions of years before we humans managed to weave our first fishing nets. Insects solved complicated problems of aerodynamics and navigation several million years ago and learned if not how to tame fire, at least how to tame light—even within their own bodies.

Insects Assemble

Whether we opt to count them by individual or species, there are good grounds for claiming that insects are the most successful class of animal on the planet. Not only are there incredible numbers of individual insects, they also account for well over *half* of all known multicellular species. They come in around a million different variants. This means that you could have an "insect of the month" calendar that featured a new species every single month for more than 80,000 years!

From A to Z, insects impress with their species' richness: ants, bumblebees, cicadas, dragonflies, earwigs, fireflies, grasshoppers, honeybees, inchworms, jewel beetles, katydids, lacewings, mayflies, nits, owl moths, praying mantises, queen butterflies, rice weevils, stink bugs, termites, urania moths, velvet ants, wasps, xylophagous beetles, yellow mealworms, and zebra butterflies.

Let's do a quick thought experiment: to get an impression of how species diversity is distributed among different groups of species, imagine if all the world's known species—big and small alike—were given UN membership. It would be an awfully tight squeeze in the assembly chambers, because even if there were only a single representative for every species, that would still add up to well over 1.5 million representatives.

Let's say we distributed power and voting rights in this "United Nations of biodiversity" according to the number of species in the different species groups. That would create new and unusual patterns, predominantly because insects would dominate, comprising more than half of all votes. And that's before we consider all the other small species, such as spiders, snails, roundworms, and the like, which alone would account for a fifth of the votes. Next up, plant species of all kinds would total roughly 16 percent, broadly speaking, while known species of fungus and lichen would command around 5 percent of votes.

Where do we fit into this picture? When we look at species diversity like this, humanity doesn't amount to much. Even if we were counted along with all the rest of the world's vertebrates—with animals such as elk and mice, fish, birds, snakes, and frogs—we would still end up with a minuscule share of power, constituting a mere 3 percent of known species diversity. In other words, we humans are totally dependent on a host of tiny species, a significant proportion of which are insects.

Dwarf Fairies and Biblical Giants

Insects come in all shapes and hues, spanning a range of sizes that is barely matched in any other class of animals. The world's tiniest insects, fairy wasps, live out the whole of their larval existence inside the eggs of other insects, which gives you a good idea of just how small they are. One of them, the teeny *Kikiki huna* wasp, is so tiny at 0.16 millimeter that you can't even see it. It takes its name from the Polynesian language spoken on Hawaii, one of the places where it is found. Logically enough, it means something like "tiny dot."

A sister species among the dwarf wasps has an even prettier name. *Tinkerbella nana* takes its genus name from the fairy in *Peter Pan*, while the species name, *nana*, is a pun referring to both *nanos*, the Greek word for "dwarf," and Nana, the name of the dog in *Peter Pan*. The Tinkerbell wasp is so small that it can land on the tip of a human hair.

It's a giant step from there to our biggest insects. There are several rivals for this title, depending on what "biggest" means. If we're talking longest, the winner is the Chinese stick insect *Phryganistria chinensis* Zhao: at 24.5 inches, it is longer than your forearm—but no thicker than an index finger. The subspecies was named for the entomologist Zhao Li, who spent six years of his life hunting down the super-stick insect after a tip from locals in the Guangxi region of southern China.

But if we're talking about the heaviest insect, the Goliath beetle is well placed. The larvae of this African giant can weigh up to 3.5 ounces, roughly the same as a blackbird. The beetle was named after Goliath, the ten-foot-tall giant of biblical fame who struck terror into the hearts of the Israelites but was nonetheless slain by a stripling named David, aided only by a sling.

The Very First Insects Predate the Dinosaurs

Insects have been around for a long time, infinitely longer than us humans. It's difficult to get a proper grasp on deep time: eons and eras, millions and billions of years. So perhaps it won't mean all that much if I say that the first insects saw the light of day around 479 million years ago. Maybe it's more helpful to point out that insects saw the dinosaurs come and go, by a long margin.

Once upon a time, long, long ago, the first plants and animals emerged from the sea and onto dry land. It was a revolution for life on Earth. Imagine, like Shaw in his book *Planet of the Bugs*, if we could have filmed that fateful moment—what an iconic video clip that would be: "One small step for bugs, one giant leap for life on Earth." Unfortunately, we'll have to settle for tracking the entrepreneurs of the insect world using fossils and our own fertile imagination.

Think back to the earth's earliest days. A few million years have passed since the first adventurous bugs poked their heads out of the sea and decided to check out new, drier neighborhoods. We are in the Devonian period, somewhat anonymously sandwiched between two better-known eras, the Cambro-Silurian period (consisting of the Cambrian, Ordovician, and Silurian periods) and the Carboniferous period (the very basis of our oil-addicted society, with all its attendant wealth and climate change). Evolution has shifted into top gear, and the first insect is now a fact: down there on the ground amid the bracken and the plants shaped like crow's feet shuffles a tiny six-legged creature with three body segments and two small antennae. It is the planet's first-ever insect, taking the first tiny steps toward total world domination by its kind.

The close interaction between insects and other life forms was crucial from their very first day on dry land. Land plants improved the life chances of insects and other bugs by providing them with sustenance up there on the stony, barren earth. In return, the bugs improved the plants' life chances by recycling the nutrition in dead plant tissue and creating soil for new growth.

The Wonder of Wings

One important reason for insects' enormous success is that they can fly. What a fantastic innovation that must have been sometime around 400 million years ago! Now insects had access to something unique: equipped with wings, they could reach the nutrition in the plants more efficiently while simultaneously avoiding earthbound enemies. For the more adventurous, wings offered brand-new opportunities to disperse to new pastures. Access to airspace also influenced choice of partner, giving insects undreamed-of opportunities to flaunt their best features in new sky-high pickup joints.

We don't know exactly when wings first developed. Perhaps they evolved from outgrowths on the thoracic area that may have served as solar collectors or a means of stabilizing the body after a jump or a fall. Perhaps the wings evolved from gills. Regardless, the most important point is that insects discovered that those gadgets of theirs were perfect for gliding down from trees or high plants. Insects with well-developed wing nubs got more food, lived longer, and as a result had more offspring, which, in turn, inherited the super–wing nubs. In this way, evolution ensured that wings became commonplace, and at a pretty rapid rate, too, in the context of geological timescales. Soon the air was alive with all manner of shimmering, whirring wings.

One point is crucial to understanding how wildly successful wings were for the early insects: *nothing* else could fly! There were not yet any birds, bats, or pterosaurs, and they would be a long time coming. That meant that insects had global dominance of the air for more than 150 million years. In comparison, our own species, *Homo sapiens*, has spent a total of just 200,000 short years on the planet.

Insects have survived five rounds of mass extinction. The dinosaurs first staggered out into the world after the third of those, around 240 million years ago. So next time you catch yourself thinking how irritating an insect is, bear in mind that this animal class has been on the planet since long before the dinosaurs. That alone merits a little respect, if you ask me.

1

Small Creatures, Smart Design

Insect Anatomy

So how are they put together, these tiny creatures we share the planet with? The following section is a crash course in insect construction. It also shows that, despite their modest size, insects can count, teach, and recognize both one another and us.

Six Legs, Four Wings, Two Antennae

What exactly is an insect? If you're in any doubt, a good rule of thumb is to start by counting legs—because most insects have six legs, all attached to the midsection of their body.

The next step is to check whether the bug has wings. These are also on the midsection. Most insects have two pairs of wings: forewings and hind wings. You have now indirectly grasped one crucial hallmark of insects: their bodies are divided into three parts. As one of many representatives of the Euarthropoda phylum, insects are formed of many segments, although in their case, these have merged into three pretty clear and distinct sections: head, thorax, and abdomen.

The old segments still appear as indentations or marks on the surface of many insects, as if somebody had cut them with a sharp implement—and in fact, that is what gave this class of animals its name: the word *insect* comes from the Latin verb *insecare*, meaning "to cut into."

The front segment, the head, isn't so unlike our own: it has both a mouth and the most important sense organs—eyes and antennae. Though insects never have more than two antennae, their eyes can vary in number and type. And just so you know: insects don't necessarily have eyes just on their head. One species of swallowtail butterflies has eyes on its penis! These help the male to position himself correctly during mating. The female also has eyes on her rear end, which she uses to check that she is laying her eggs in the right place.

If the head is the insects' sensory center, the midsection—the thorax—is the transport center. This segment is dominated by the muscles needed to power the wings and legs. It is worth noting that, unlike those of all other animals that can fly or glide—birds, bats, flying squirrels, flying fish—insects' wings are not repurposed arms or legs but separate motor devices that supplement the legs.

The abdomen, which is often the largest segment, is responsible for reproduction and also contains most of the insect's gut system. Gut waste is excreted at the rear. Usually. The minute gall wasp larvae, which live out their larval existence in the gall a plant builds around them, are extremely well brought up. They know it's wrong to foul your own nest, and since they are trapped in a one-room apartment without a toilet, they have no choice but to hold it in. Only after the larval stage is complete are the gut and the gut opening connected.

Living in an Invertebrate World

Insects are invertebrates—in other words, animals without a backbone, skeleton, or other bones. Their "skeleton" is on the outside: a hard yet light exoskeleton protects the soft interior against collision and other external stresses. The outermost surface is covered in a layer of wax, which offers protection against every insect's greatest fear: dehydration. Despite their small size, insects have a large surface area relative to their tiny volume, meaning that they are at high risk of losing precious water molecules through evaporation, which would leave them as dead as dried fish. The wax layer is a crucial means of hanging on to every molecule of moisture.

The same material that forms the skeleton around the body also protects the legs and wings. The legs are strong, hollow tubes with a number of joints that enable the insect to run, jump, and do other fun things.

But there are a few disadvantages to having your skeleton on the outside. How are you supposed to grow and expand if you're shut in like this? Imagine bread dough encased in medieval armor, expanding and rising until it has nowhere left to go. But insects have a solution: new armor, soft to start with, forms beneath the old. The old, rigid armor cracks open, and the insect jumps out of its skin as casually as we'd shrug off a shirt. Now it's crucial that it literally inflate itself to make the new, soft armor as big as possible before it dries and hardens. Because once the new exoskeleton has hardened, the insect's potential for growth is fixed until another molting paves the way for new opportunities.

If you think this sounds tiring, it may be a consolation to hear that (with a few exceptions) the lengthy molting process occurs only in insects' early lives.

A Time of Transformation

Insects come in two variants: those that change gradually through a series of moltings and those that undergo an abrupt change in the process of developing from child to adult. These transformations are called *metamorphosis*.

The first type—e.g., dragonflies, grasshoppers, cockroaches, and true bugs (the order Hemiptera)—gradually change in appearance as they grow, a bit like us humans, except that we don't have to shed our entire skin in order to grow. For these insects, the childhood stage is known as the nymph stage. The nymph grows, casts off its exoskeleton a few times (just how many varies by species, but often three to eight times), and becomes increasingly like the adult version. Then, finally, the nymph molts one last time and crawls out of its used larval skin equipped with functioning wings and sex organs. Voilà! It has become an adult!

Other insects undergo a complete metamorphosis—an almost magical change in appearance from child to adult. In our human world, we have to turn to fairy tales and fantasy for examples of this sort of shape-shifting, such as kissed frogs turning into princes or Minerva McGonagall shape-shifting into a cat. But for insects, kissing and spells aren't the cause of the change. The metamorphosis is driven by hormones and marks the transition from child to adult. First the egg hatches into a larva that looks nothing like the creature it will ultimately become. The larva often looks like a dull, pale, rectangular bag, with a mouth at one end and an anus at the other (although there are some exceptions, including many butterflies). The larva molts several times, growing bigger on each occasion but otherwise looking pretty much unchanged.

The magic happens in the pupal stage—a period of rest in which the insect undergoes the miraculous change from anonymous "bag creature" to an incredibly complicated, ingeniously constructed adult individual. Inside the pupal case, the whole insect is rebuilt, like a Lego model whose blocks are pulled apart and put back together again to make an entirely different shape. In the end, the pupa splits and out climbs "a beautiful butterfly"—as described in one of my all-time favorite children's books, *The Very Hungry Caterpillar*. Total transformation is brilliant and undoubtedly the most successful variant. Most insect species on the planet, 85 percent of them, undergo this type of complete metamorphosis. This includes the dominant insect groups, such as beetles, wasps and their relatives, butterflies and moths, and flies and mosquitoes.

The ingenious part is that they can exploit two different diets and habitats as child and adult, concentrating on their central task in each phase of their lives. The earthbound larvae, whose focal point is energy storage, can be eating machines. Then, in the pupal stage, the accumulated energy is melted down and reinvested in a new organism: a flying creature dedicated to reproduction.

The connection between larvae and adult insects has been known since ancient Egyptian times, but people didn't understand what was happening. Some thought that the larva was a stray fetus that eventually came to its senses and crawled back into its egg—in the form of the pupa—in order to be born at last. Others claimed that two totally different individuals were involved, the first of which died and was then resurrected in a new form.

Only in the 1600s did Jan Swammerdam and his microscope demonstrate that the larva and the adult insect were the same individual throughout. The microscope enabled people to see that if a larva or pupa was carefully cut open, clearly recognizable elements of the grown insect could be found beneath the surface. Swammerdam enjoyed displaying his skills with scalpel and microscope before an audience and used to demonstrate how he could remove the skin from a big silkworm larva and reveal the wing structure beneath, complete with the characteristic veined patterning on the wings.

Even so, this did not become general knowledge until much, much later. In his journal, Charles Darwin noted that a German scientist was charged with heresy in Chile as late as the 1830s because he could transform larvae into butterflies. Experts still discuss the exact details of the metamorphosis process even now. Luckily, there are still some mysteries left in the world!

Breathing through a Drinking Straw

Insects don't have lungs and don't breathe through their mouths as we do. Instead, they breathe through holes in the sides of their bodies. The holes run, like drinking straws, from the surface of the insect into its interior, branching out along the way. Air fills the straws, and the oxygen then passes out of them and into the body's cells. This means that the insects don't need to use their blood to transport oxygen to the various nooks and crannies of their bodies. However, they do still need some kind of blood—known as *hemolymph*—to carry nutrition and hormones to the cells and to clear them of waste material. Since insect blood doesn't transport oxygen, there is no need for the ferrous red substance that colors our mammal blood red. Consequently, insect blood is colorless, yellow, or green. That is why your car windshield doesn't end up looking like a scene from a bad crime novel when you're driving along on a hot, still summer afternoon but ends up covered in yellowish green splatters instead.

Insects don't even have veins and arteries: instead, insect blood sloshes around freely among the bodily organs, down into the legs, and out into the wings. To ensure a bit of circulation, there is a heart of sorts: a long dorsal tube with muscles and openings on the side and at the front. Muscle contractions pump the blood forward from the rear, toward the head and brain.

Insects' sensory impressions are processed in the brain. It is tremendously important for them to pick up signals from their surroundings in the forms of scent, sound, and sight if they are to find food, avoid enemies, and pick up mates. Although insects have the same basic senses as we do—they sense light, sound, and smell and can taste and feel—most of their sense organs are constructed in a totally different way. Let's take a look at insects' sensory apparatus.

The Fragrant Language of Insects

The sense of smell is important for many insects, although they lack a nose, doing most of their smelling through their antennae instead. Some insects, including certain male moths, have large, feathery antennae that can pick up the scent of a female several miles away even in extremely low concentrations.

In many ways, insects speak through smell. Scent molecules allow them to send each other various kinds of messages, from soppy personal ads such as "Lonesome lady seeks handsome fella for good times" to ant restaurant recommendations: "Follow this scent trail to a delicious dollop of jam on the kitchen counter."

Spruce bark beetles, for example, don't need Snapchat or Messenger to tell each other where the party is. When they discover an ailing spruce tree, they shout about it in the language of scent. This enables them to gather together enough beetles to overpower a sickly living tree—which then ends its days as a kindergarten for thousands of beetle babies.

We miss out on most of these insect scents, which we simply can't smell. But if you wander beneath the greenery of ancient trees on a late-summer day in the town of Tønsberg, southern Norway, you may be lucky enough to pick up the most delightful aroma of peaches: it is the scent of the hermit beetle, one of Europe's largest and rarest beetles, wooing a girlfriend in the neighboring tree. The substance it uses rejoices in the thoroughly unromantic name of gamma-decalactone, and we humans produce it in labs for use in cosmetics and to add aroma to food and drinks.

The scent is very helpful to the hermit beetle, which is heavy and sluggish and seldom flies, or not far at any rate. It lives in ancient hollow trees, where its larvae gnaw on rotted wood debris, and it's a real homebody: a Swedish study found that most adult hermit beetles were still living in the same tree they were born in. This lack of interest in travel complicates the business of finding new hollow trees to move into, and the situation is hardly helped by the fact that old, hollow trees are an unusual phenomenon in today's intensively exploited forests and farmlands. As a result, the species, which is scattered across western Europe from southern Sweden to northern Spain (though not the British Isles), is decreasing all over its range and is protected in many European countries. In Norway, it is considered critically endangered and can be found in only one place: an old churchyard in Tønsberg. Or two places, to be precise, because some individuals have recently been moved to a nearby oak grove in an effort to secure the survival of the species.

Flowery Temptresses

Flowers have realized that scent is important to insects—or rather, millions of years of mutual evolution have resulted in the most incredible interactions. The world's largest flower, which belongs to the *Rafflesia* genus and is found in Southeast Asia, is pollinated by blowflies. This means that "a scent of warm summer sun meets a cool evening breeze with a hint of amber and sensual vanilla"—to borrow perfume industry terminology—isn't going to cut the mustard. No, indeed. If you want blowflies to come visiting, you need to yell at them in blowfly language. That is why the world's biggest flower smells like a dead animal whose carcass has been lying around in the heat of the jungle a couple of days too many—a stench of rotting flesh that is irresistible if you happen to be a blowfly.

But you don't have to go the jungle to find examples of flowers that speak the insects' language of scent. The fly orchid is a protected native European species, rare in Norway and the United Kingdom but widespread throughout central Europe. It has strange brownish blue flowers that look just like the female of a certain digger wasp species. And its beautiful appearance is supplemented by the right scent: the flower smells identical to a female digger wasp on the prowl. So what is a bewildered newly hatched male digger wasp, whose short life is dominated by a single thought, to do? He falls for the trick and tries to mate with the flower. When things don't go so well, he moves on to what he thinks is the next female and tries again. No luck there, either. What he doesn't know is that during these ill-fated pairings, he has picked up some yellow structures that contain the fly orchid's pollen, so the male digger wasp's feverish flirting contributes to the flowers' pollination.

And if you're concerned about the fate of the unfortunate male, please don't despair. The real females hatch a few days after the males, and then things really start heating up. In this way, the existence of both the fly orchid and the digger wasp is ensured.

Ears on Their Knees and Deathwatch Beetles

Although communicating through scent is important for insects, especially if they're searching for a mate, some rely on sound to find a partner instead. The grasshopper's song is not designed to create the sound of summer for us humans but to find a girlfriend for the little green creature; it is usually the male calling out to the female, in the same way as male birds are frequently the keenest warblers. If you've heard the deafening wall of sound cicadas create in southern climes, bear in mind that it would be twice as loud if the ladies joined in. But as an ancient Greek saying has it, "Blessed are the cicadas, for they have voiceless wives." Controversial as we may find this statement in modern society, let me just add that it may be pretty smart of the females to keep their lips zipped. Lovesick fellow cicadas aren't the only ones attracted by the song. Scary parasites lie in wait listening, then sneak up to lay a tiny egg on the soloist. And although it might look quite innocent, it's game over for the singer. The egg hatches into a hungry larva, which eats up the cicada from the inside out. Enough said.

Insects have ears in all sorts of peculiar places but rarely on their heads. They may be on their legs, their wings, their thorax, or their abdomen. Certain moths even have ears in their mouths! Insect ears come in a number of variants, and even though all of them are XXXS size, some of them are incredibly intricate. One type has a vibrating membrane, like a tiny drum, whose skin is set into motion every time sound waves from the air reach it. It isn't unlike our own inner ear, just in a simplified miniature version.

Insects can also sense sound through different sensors connected to small hairs that pick up vibrations. Mosquitoes and fruit flies have these kinds of sensors on their antennae, while the bodies of butterfly larvae may be covered in sensory hairs, which they use to hear, touch, and taste. Some ears can pick up sounds a long way off, while others operate only over very short distances. It's sometimes difficult to say what "hearing" actually is: for example, are you hearing or feeling when you pick up vibrations in the stem of the stalk of grass you're perched on?

If you are small, you can use an amplifier to boost your sound—as do several species among the wood-boring beetles (*Hadrobregmus pertinax* and *Xestobium rufovillosum*). In the olden days, people thought that the sound they made was a forewarning of imminent death, but the actual explanation is much more prosaic. These beetles live out their larval existence in rotting woodwork, often in the timbers of houses. As adults, the beetles find partners by banging their heads against the wall. The sound transmits effectively through the woodwork and is picked up by both the beetles and us humans. This repetitive knocking is reminiscent of a ticking clock, or perhaps even more like somebody drumming his fingers impatiently on a table. According to ancient superstition, these sounds meant that somebody would soon die: they were a clock counting down the person's final hours or the Grim Reaper waiting restlessly. It was probably just easier for people to hear the sounds at night in a quiet house, when, say, they were keeping watch at somebody's deathbed.

Fiddling on the World's Tiniest Violin

There are other insect sounds that we hear distinctly even in the clear light of day, such as the cicadas' song. Even so, cicadas aren't the winners in the competition for the world's noisiest insects. Adjusted for size, an aquatic insect a mere 2 millimeters long is the one most likely to walk away with the prize, because the male water boatman, part of the Micronectidae family, competes for the ladies' attention by making music. But how are you supposed to serenade your sweetheart when you're the size of coarsely ground pepper? Well, the little water boatman does it by playing himself, using his abdomen as a string and his penis as a bow.

Several years ago, a team of French scientists set up underwater microphones to record the song of male water boatmen—the first-ever bootleg recording of this serenade. And what a hit it was in its way: the scientists believed they could prove that these tiny creatures with their fiddling penises exceeded all bounds of reason when it came to sound production. An average sound level of no less than 79 decibels made by a critter a mere 2 millimeters in length: on land that's equivalent to the sound of a freight train passing by at a distance of around 50 feet.

It seems almost beyond the realm of possibility, and it may actually not be true, either, because it's a complicated business to compare sounds underwater and in the air. Perhaps the water boatman will turn out not to be the world's noisiest insect after all. But the fact that it fiddles with its own penis—well, you can't take *that* away from it.

Tongues Beneath Their Feet

Imagine if you could walk barefoot through the forest in the summertime and actually *taste* the blueberries in the bushes as you stepped on them! This is what houseflies do. They taste with their feet. And flies are unbelievably hypersensitive, apparently a hundred times as sensitive to sugar as we are with our tongues.

But there are a few disadvantages to being a fly, on top of being a mostly unwanted creature to begin with. Flies don't have teeth or any other equipment that would enable them to eat solid food, which dooms them to an eternally liquid diet. So what is a poor housefly to do when it lands on something tasty, like your slice of bread? Well, it uses digestive enzymes from its belly to turn the food into a smoothie. To do so, it has to regurgitate some of its gastric juices onto the food, which isn't so great for us because it means that bacteria from the fly's last meal—possibly far from anything we'd classify as food—may end up on our slice of bread. But it's great for the fly, which can now suck up the food. The housefly's mouth is like a spongy vacuum cleaner head on a short shaft. The whole thing is attached to a kind of pump in the head, which creates suction, allowing the fly to vacuum up the yummy, nutritious soup.

Houseflies' poor table manners and somewhat varied diet, which includes items such as animal dung, are the reasons they spread infection. The flies aren't dangerous in themselves, but, like used syringes, they can carry infections and pass them on to us.

And now that I think about it, maybe it's just as well we humans taste with our tongue and not our feet. Blueberry shrubs are one thing, but the thought of going around tasting the insides of your shoes all winter long is hardly appealing.

A Multifaceted Life

Insects' senses are adapted to their environment and needs. Whereas dragonflies and flies need good vision, insects that live in caves may be totally blind. Insects that come into close contact with flowers, such as honeybees, can also see colors, but their color spectrum is shifted upward, so they don't see red light. On the other hand, unlike us humans, they can see ultraviolet light. This means that many flowers we see as monotone, such as sunflowers, have distinctive patterns for a bee, often in the form of "landing strips" that direct them toward the source of nectar in the flower.

Insects' compound eyes consist of many individual eyes. The brain merges all the tiny pictures together into a single large image, although it is coarser and fuzzier than the way we see the world. It looks a bit like a low-res photograph on your computer screen when you've zoomed in too close. There are plenty of reasons why insects don't have driving licenses, of course, but sight is a big one: they would never be able to read a road sign at 60 feet, as the image would be too blurry.

That said, their vision is supremely adapted to the tasks that will fill their days. Take whirligig beetles, for example, shiny black pearls of beetles that dash around on the surface of the waters of our lakes. They have two pairs of eyes with different refractions: one pair for seeing clearly under water so that they can watch out for hungry perch, the other for seeing clearly above the water so that they can find food on the surface.

Insects can also see a property we humans are blind to: polarized light. This has to do with which plane the light is oscillating in, and it alters when sunlight is reflected—in the atmosphere or off a shining surface such as water. But let's go easy on the physics and restrict ourselves to saying that insects use polarized light as a compass that enables them to orient themselves. We humans relate to polarized light only when we put on a pair of Polaroid sunglasses to reduce the glare of reflected light.

In addition to having compound eyes, insects may have separate simple eyes whose main function is to distinguish between light and dark. Next time you meet a stinging wasp, look it deep in the eyes and note how, in addition to the compound eyes on either side of its head, it has three simple eyes in a neat triangle on its forehead.

The World's Most Skillful Hunter Sees You and You and You . . .

When it comes to having eyesight adapted to their daily business, dragonflies are in a class of their own: their vision is a major reason why these insects are deemed to be among the world's most efficient predators.

Lions may put on an impressive display when they're hunting in a pride, but they manage to chase down their prey only one in every four times. Even the great white shark, with its terrifying three-hundred-toothed grin, fails in half of all its attempted attacks. The dragonfly, however, excels as a lethal hunter, succeeding in more than 95 percent of its attempts.

Part of the reason dragonflies are such skilled hunters is their extraordinary command of the skies. Their four wings can move independently of one another, which is unusual in the insect world. Each wing is powered by several sets of muscles, which adjust frequency and direction. This enables a dragonfly to fly both backward and upside down and to switch from hovering motionless in the air to speeding off at a maximum of close to 30 miles an hour. No wonder the US army uses them as models when designing new drones.

But their vision also makes a significant contribution to their success. And perhaps it is hardly surprising that they have good eyesight when almost their entire head consists of eyes. In fact, each eye is made up of 30,000 small eyes, which can see both ultraviolet and polarized light as well as colors. And since the eyes are like balls, the dragonfly can see most of what is happening on all sides of its body.

Its brain is also prepped for supersight. When we humans see a rapid sequence of images, we see them in a flowing movement, a film, if there are more than around twenty images per second. However, a dragonfly can see up to three hundred separate images per second and interpret every one of them. In other words, a movie ticket would be wasted on a dragonfly. Where you and I see a moving film, it would simply see a very rapid slide show.

The dragonfly brain is also capable of focusing over time on one specific section of the enormous quantity of visual impressions being received. It has a kind of selective attention that is unknown among other insects. Imagine that you're traveling across the sea in a boat and see another boat ahead of you, at a given angle to you. If you ensure that you always have the boat at exactly the same angle in your field of vision, you will end up meeting. In a somewhat similar way, the dragonfly brain can lock its attention on approaching prey, coordinating its speed and direction to ensure a strike—and yet another successful hunt. Intricate, well-designed sensory organs alone are not enough: you also need a brain that can process all the information as it streams in, seeking out relevant patterns and connections and sending the correct messages out again to different parts of the body. And even though insects have tiny brains, we will see that they are a lot smarter than we might assume.

Go to the Ant and Be Wise

Carl Linnaeus, the great Swedish biologist who classified our species, placed insects in a separate group, in part because he believed they didn't have brains at all. Maybe that's not such a surprise, because if you behead a fruit fly, it can live pretty much as normal for several days, flying, walking, and mating. Eventually, of course, it will starve to death, because having no mouth means not eating. The reason insects can survive in a headless state is that they have not only a main brain in their head but also a nerve cord that runs through their entire body, with "minibrains" in each joint. Consequently, many functions can be performed regardless of whether or not the head is in place.

Are insects intelligent? That depends on what you mean by intelligence. According to Mensa, intelligence is the "ability to acquire and analyze information." Now, it's unlikely that anybody's going to argue that insects deserve to be Mensa members, but the fact is that they never cease to surprise us with their ability to learn and make judgments. Some things we believed to be the sole preserve of large vertebrates with proper brains also turn out to be within the capabilities of our tiny friends.

But not all insects are created equal, and there are great differences among them. Those with dull lives and simple habits are the least bright. You don't need the wisdom of Solomon if you're going to spend most of your life snugly tucked up in an animal hide with your sucking snout stuck in a vein. However, if you're a honeybee, a wasp, or an ant, you're more in need of intelligence. The cleverest insects are the ones that look for food in lots of different places and also form close bonds with one another; in other words, the ones that live alongside many others in a community. These critters must constantly be making judgments: Is that yellow thing over there a flower with sweet nectar, or is it a hungry crab spider? Will I be able to carry that conifer needle alone, or will it take several of us? Do I need to take a sip of this nectar to keep myself going, or should I take it home to Mom?

The social insects divide up jobs, share experiences, and "talk to each other" in an advanced way. This requires capacity for thought. To cite Charles Darwin: "The brain of an ant is one of the most marvelous atoms of matter in the world, perhaps more so than the brain of a man." And that was without knowing what we now know: that ants are capable of teaching skills to other ants.

The ability to *teach* has long been seen as exclusive to us humans, almost a proof of an advanced society. Three quite specific criteria distinguish teaching from other communications: it must be an activity that happens only when a teacher meets an "ignorant" pupil; it must involve a cost for the teacher; and it must make the pupil learn more rapidly than it otherwise would have. The term is used for communication about concepts and strategies, so the honeybees' dance (see page 22), which is more about process, is generally not viewed as teaching. However, it turns out that ants are capable of teaching things to other ants, through a process known as *tandem running*, in which an experienced ant shows the way to food. This occurs in a European species, *Temnothorax albipennis*, which relies on landmarks such as trees, stones, and other things, as well as scent trails, to remember the way from the anthill to a new source of food. In order for several ants to be able to find the food, a she-ant (all worker ants are female) who knows the way must teach it to the others. The teacher runs on ahead to show the way but constantly stops to wait for the pupil, who runs more slowly, apparently because she needs time to take note of the landmarks they are passing. When the pupil is ready again, she touches the teacher with her antennae and they continue on their journey. The behavior therefore satisfies the three criteria of genuine teaching: There's a teacher and an "ignorant" pupil involved, the teacher must stop and wait, so there's a cost, and the pupil must learn her way to the food more quickly than she would have on her own.

Bumblebees have also recently been inducted into the exclusive little group of animals that can teach tricks to their peers. Swedish and Australian scientists successfully trained bumblebees to pull on a string to gain access to nectar. They made artificial blue flowers in the form of plastic discs, which they filled with sugar water. When these were covered with a transparent plate of plexiglass, the only way of gaining access to the sugar water was to pull on a string attached to the fake flower. If the scientists let only untrained bumblebees loose on the covered flowers, they didn't understand a thing. None of them pulled on the string—a great starting point. Then the bumblebees were given a chance to get acquainted with the "flowers," to learn about the reward they offered. Gradually, the fake flowers were pushed farther and farther beneath the transparent plexiglass plate. This time, when the fake flowers were finally pushed fully beneath the plate, twenty-three out of forty bumblebees began to pull on the string. In this way, they drew the fake flowers out and were able to suck up the sugar water. Admittedly it was a long lesson: the whole business took a good five hours of training per bumblebee.

The next step was to see whether these trained bumblebees

could teach others their peculiar trick. Three bumblebees were selected as "teachers." New, untrained bumblebees were placed with them in a small transparent cage close to the flowers to watch and learn. Fifteen of the twenty-five "pupils" grasped the point by watching how their teacher did it and themselves managed to pull out their reward when they got to try afterward. All in all, this experiment showed both that the bumblebees could learn this rather unnatural skill and that they were capable of teaching the strategy to others.

Clever Horse Hans and the Even Cleverer Bee

Hans the Horse of Germany was a global celebrity in the early 1900s. He couldn't just count, he could also calculate—or so people thought. The horse could add, subtract, multiply, and divide. He answered math problems by banging out the correct answer with his foreleg, and the horse's owner, math teacher Wilhelm von Osten, was convinced that the animal was just as clever as he was. In the end, it turned out that Hans couldn't calculate at all or even count. That said, he *was* a whiz at reading the minuscule signals in his questioner's body language and facial expressions. The person setting the problem had to count, too, to make sure that Hans was giving the right answer, and a tiny, unconscious signal he made when the horse reached the correct number was all that Hans needed. In fact not even the psychologist who eventually unmasked Hans was able to control those signals.

However, bees actually can count—not very far, and they are no more capable of the four types of calculation than Hans was.

Even so, it's a pretty impressive feat for a creature with a brain the size of a sesame seed. To measure this ability, honeybees were placed in a tunnel and trained to expect a reward after passing a certain number of landmarks, regardless of how far they had to fly. It turned out that they could count up to four, and once they had learned to do that, they were able to count the landmarks even if they were a new type they had never seen before.

And bees aren't just good at math (well, considering their size); they are also good at languages.

Dancing Bee People

At around the same time as Osten and his not-so-clever horse were alive, a future Nobel Prize winner was growing up in the neighboring country of Austria. Even as a child Karl von Frisch loved animals, and his mother must have been extremely tolerant, since she put up with the abundant array of wild animals he brought home as pets. Over the course of his childhood, he noted 129 different pets in his journal, including 16 birds; 20-odd types of lizards, snakes, and frogs; and 27 different fish. Later, as a zoologist, he was especially interested in fish and their color vision. But almost by chance—largely because his aquatic research subjects displayed an unfortunate tendency to expire on the way to the conferences where he was supposed to be demonstrating his experiments—he switched to studying bees instead.

Frisch made two major discoveries: he proved that bees can see colors and that they can tell each other where to find food by performing a sophisticated dance. That was what won him a Nobel Prize in 1973. Frisch showed that when a honeybee finds a rich source of nectar, she returns home to the others and tells them where the flowers are. She dances in a kind of figure eight, waggling her rear and vibrating her wings in the parts of the dance when she is moving in a straight line. The speed of the dance communicates the distance to the flowers, while the direction she dances in, in relation to a vertical line, describes where the flowers are relative to the position of the sun.

Today, bees' dance language is one of the best-researched and best-mapped examples of animal communication. But history could have turned out quite differently: in Adolf Hitler's Germany, this research was nearly brought to a halt when it had barely begun. In the 1930s, when Frisch was working at the University of Munich, Hitler sympathizers scoured the university's employee roster to root out Jewish workers. When Frisch's maternal grandmother proved to have been Jewish, he was fired. But he was rescued by a tiny parasite—one that caused a disease in bees that was in the process of wiping out Germany's bee population. Beekeepers and colleagues managed to persuade the Nazi leadership that Frisch's future research was crucial if German beekeeping were to be rescued. The country was at war and in dire need of all and any foodstuffs farming could produce. A collapse of the honeybee population was unthinkable. Thus Frisch was able to carry on his research, for the good of both bee knowledge and Frisch's career.

I've Seen That Face Before

For a long time, we believed that only mammals and birds were capable of distinguishing among individuals, the very foundation of the capacity to develop personal relationships. This belief persisted until an inquiring scientist, with the help of some model airplane paints, began face painting wasps. The species concerned was *Polistes fuscatus*, an American member of the family of paper wasps. Paper wasps build nests from chewed-up wood fiber that look like a rosette of small larval cells. The nest hangs on a stalk, like an upside-down umbrella. Unlike regular stinging wasps, which also build nests from wood pulp, paper wasps' nests do not have a protective envelope around the comb of larval cells.

This wasp lives in a strictly hierarchical society, where it's crucial to know who's the boss. Maybe that's why they're so good with faces. A wasp whose face had been painted in a way that altered her pattern of stripes met with an aggressive reception from her fellow inhabitants when she returned to the nest. They didn't recognize her and were confused. As a control, the scientists also painted other wasps without altering their patterns of stripes. Those wasps did not experience any reactions on their return to the nest.

Another fascinating point is that after a few hours of jostling, the other inhabitants of the nest got used to the face-painted wasp's new look. The aggression diminished, and everything went back to normal. The other wasps had learned that this was indeed the same old Waspella, despite her makeover. This implies that wasps actually have the capacity to recognize and distinguish among individual members of their community by their detailed facial cues or "features."

Honeybees take the whole business up a few notches: they can distinguish among human faces in the form of photographic portraits. What's more, they can remember a face they've become familiar with for at least two days. It is doubtful whether the bees relate to *what* they are seeing. They seem to believe the portraits they are presented with are really funny flowers, with the darker areas of eyes and mouth representing recognizable patterns on "petals" that are actually the outline of the portrayed face.

This is new and exciting information, which forces us to rethink how facial recognition works: after all, we're saying that an animal whose brain is smaller than the letter *o* in this book is able to achieve similar things as we humans, with our cauliflower-sized brain boxes. Greater understanding of these processes may be able to help people who suffer from face blindness (prosopagnosia), which is the inability to recognize faces.

Perhaps this knowledge could be used in surveillance, at airports, say. Not by installing a glass cage of buzzing bees to scrutinize us sternly as we go through customs (although that would be pretty cool!) but rather by translating the principles that enable bees to recognize facial patterns into a logic that computers can follow. One hope is that this could lead to improved automatic facial recognition—of, say, wanted criminals—via surveillance cameras in crowded places.

What Shall We Call the Beetle? Names and Insect Groups

In an attempt to organize the hordes of tiny creatures, we humans have split them into groups according to how closely related they are. It's an ingenious system that starts with the *kingdom*, which is then divided into *phylum* and *class*, which are again divided into *order*, *family*, and *genus* before we come to *species*.

Take the common wasp, for example. It is a species that belongs to the animal kingdom, the Euarthropoda phylum, the Insecta class, the Hymenoptera order, the stinging wasp family, the *Vespula* genus, and, finally, the common wasp species.

All species have a two-part Latin name, which is written in italics. The first part tells you which genus the species belongs to, and the second part identifies the species. This system, introduced by Carl Linnaeus in the 1700s, makes it easier for biologists to be certain that they're talking about the same species even when they're communicating across national borders and language barriers. The common wasp, for example, has been given the name *Vespula vulgaris*. You can often grasp the meaning of the Latin names: for example, *vulgaris* means "common" (and is also the origin of the word *vulgar*).

Sometimes the Latin name tells us something about the insect's appearance, as with the *Stenurella nigra* beetle, where *nigra* describes the color of this totally black species. Other times, the name has been borrowed from mythology, as in the case of the beautiful peacock butterfly, *Aglais io*. Io was one of Zeus's mistresses, who also lent her name to one of Jupiter's moons.

With hundreds of thousands of insects to name, entomologists sometimes go a bit wild, calling species after their favorite artists, such as the *Scaptia beyonceae* horsefly (see page 41) or characters from much-loved films, such as the *Polemistus chewbacca*, *P. vaderi*, and *P. yoda* wasps. Sometimes the species names contain a pun that you discover only when you say them out loud. Just try pronouncing the names of the bean-shaped beetle *Gelae baen* and *Gelae fish* or the parasitic wasps *Heerz lukenatcha* and its relative *Heerz tooya*!

Orders for Order

There are around thirty different orders of insects in the world. Beetles, wasps and their relatives, butterflies and moths, flies and gnats, and true bugs are the five largest. Other orders include dragonflies, cockroaches, termites, orthoptera (grasshoppers and crickets), caddis flies, stone flies, mayflies, thrips, lice, and fleas.

Let's start with beetles (coleoptera), one of the largest orders of insects worldwide, despite tough competition from the wasp order, where improved knowledge is leading to a steady rise in the number of species. The hallmark of beetles is that their forewings are hard, forming a protective shell over their back. Beyond that, beetles are incredibly varied in appearance and lifestyle and can be found on both land and water. There are more than 170 different beetle families, some of the largest being true weevils, scarab beetles, leaf beetles, ground beetles, rove beetles, longhorn beetles, and jewel beetles. All in all, there are around 380,000 known beetle species worldwide.

The wasp order (Hymenoptera) consists of familiar insects such as ants, bees, bumblebees, and stinging wasps, including many species that are social and live in colonies containing hordes of female workers and one or more queens. The order also encompasses many lesser-known sawflies and a huge number of parasitic wasp species. So far, we have identified more than 115,000 species in this order, but the number is rising steadily and this is probably the largest order of insects.

Butterflies and moths (of the Lepidoptera order) have wings covered in tiny scales arranged like roof tiles. There are more than 170,000 lepidopteran species in the world, but many are small and unassuming. The best known are of course butterflies, comprising around a hundred large, diurnal species that are often beautifully colored and patterned. The nocturnal species are known as moths.

Flies, or dipterans, include not only species we commonly call flies, such as blowflies and horseflies, but also mosquitoes, gnats, and crane flies. Their Latin name derives from the fact that they have only two wings (*di* means "two," *ptera* means "wing"), whereas insects normally have four, as mentioned earlier. In dipterans, the hind wings have been repurposed as small, club-shaped gadgets that help them achieve balance in flight. We know of at least 150,000 species of dipterans worldwide.

Most people are less familiar with the order of true bugs (Hemiptera), even though it encompasses more than 80,000 species. The group includes a variety of different-looking insects, such as shield bugs, stink bugs, bedbugs, pond skaters, cicadas, aphids, and scale insects. They all have beak-shaped mouths that serve as a kind of drinking straw they use to suck up their food—often sap from plants, although a few are predators or bloodsuckers. So although we commonly use the word "bug" to describe any sort of tiny creature, the true bugs are a specific group of insects. True bugs of the most familiar suborder, Heteroptera, are similar to beetles in body shape, but are identifiable by triangular marks on their backs.

And just so you know: spiders aren't insects. They belong to the same phylum, Euarthropoda, but a separate class, Arachnida, which they share with other creatures such as mites, scorpions, and daddy longlegs (known as weaving women in Norwegian because they move two of their eight legs as if they were pushing a shuttle to and fro across a loom).

Millipedes, centipedes, and wood lice aren't insects, either.

To take the simplest hallmark, they all have too many legs and belong to various other groups of invertebrates. Nor are the supercute springtails insects, despite having six legs, although they are *nearly* insects. That said, insect researchers are huge fans of a teeming multilegged community, so springtails and arachnids are often allowed into the fold when we discuss insects anyway. That is true in this book, too.

2

Six-Legged Sex

Dating, Mating, and Parenting

What accounts for the enormous success of insects as an animal group? Why are they so rich in species, and why are there so very many of them? Put simply: because they are small, supple, and sexy.

Life on our planet spans more than ten orders of magnitude—from mycoplasma bacteria (measurable in nanometers) to the gigantic redwood trees of California, which can grow to heights of more than 300 feet. Insects can be found in six of these, all at the lower end of the scale—from wingless male fairy wasps, which are smaller than the cross section of a human hair, to stick insects the length of your forearm (see page xvi). In other words, most insects are small, so they need only a tiny hiding place to avoid their enemies, and can exploit resources that are of no interest to larger animals.

Insects are also unbelievably supple, in the sense that they are flexible and adaptable. Their wings allow them to disperse over areas that are extremely large in relation to their modest size, and their mastery of the three dimensions of airspace gives them access to a great many sources of nutrition. The fact that most insects spend their childhood in a body shape totally different from their adult form (see page 4) means that they can exploit entirely different habitats and sources of food in the different stages of their lives—and the young do not compete with the adults for food.

Last but not least, insects have an astonishing capacity for reproduction. There must surely have been a fly on the wall who thought God was talking to her when he said "Be fruitful, and multiply, and replenish the earth and subdue it" (Genesis 1:28, King James version). Just consider this, as calculated by Borror and colleagues: Take two fruit flies and leave them in ideal living conditions for a year—equivalent to twenty-five generations for fruit flies. Every fruit fly mother lays a hundred eggs. Let's say that they all grow to adulthood and that half of them are females that mate and lay another hundred eggs. Once the year is over, you will be left with the twenty-fifth generation, and that alone will amount to almost a tredecillion sweet little red-eyed fruit flies. A tredecillion is the figure 1 followed by forty-two zeros. To make this figure more meaningful, imagine packing these flies together as densely as possible into one enormous fruit fly ball. You'd end up with a sphere whose diameter exceeded the distance between the earth and the sun! It's a good thing these insects have so many enemies, because otherwise there wouldn't be any room left on Earth for us humans.

Fortunately (we might say) most of the insect eggs will never catch so much as a glimpse of adult life. Most insects starve to death, are eaten, or die in some other way long before they start a family. It's a harsh struggle. Over time, an incredible range of adaptations has come about, especially when it comes to choice of partner and propagation. We will look at some of them in this chapter.

Fifty Shades of Strange

Insects' senses are crucial to their quest to find a partner, and the competition is tough. But the struggle is far from over when boy meets girl. On the contrary: it has barely begun—because the question of how to pass on as much of their genetic material as possible may have different answers for the two sexes.

For example, it isn't unusual for the female to mate with several males over a short period, and the male takes a dim view of this because it means that his sperm face competition. Consequently, many insects come equipped with a male sex organ reminiscent of a Swiss Army knife, complete with all sorts of imaginatively shaped scrapers, ladles, and spoons. The purpose? To eliminate any sperm that got there before their own.

This tool kit also comes in handy if the previous male has resorted to another trick: plugging up the female's genital opening. The idea is to create a kind of homemade chastity belt that prevents the female from being able to mate again. The gambit is only partially effective, as male number two simply uses his scrapers, pike poles, and hooks to remove the bung and gain access for his own equipment. So much for candlelight and tender caresses!

Another trick the male deploys is to ensure that as much of his sperm as possible is delivered to the female and that she has as little time as possible available for other males. He does this by making their mating process as lengthy as possible. Some species take this to extremes: the Southern green stink bug, *Nezara viridula*, which has spread all over the world, including to the United Kingdom as a stowaway on imported foodstuffs, can keep at it for ten whole days. And that's still nothing compared with Indian stick insects, which have apparently been known to stay stuck together for an insane seventy-nine days in an extreme-sport version of tantric sex!

As well as engaging in prolonged mating, the male often keeps an eye on the female postconsummation. Have you ever seen those small blue damselflies, a close relative of dragonflies, perching or flying around in pairs? Sometimes the coupling creatures resemble a heart—although that doesn't imply any human notion of romance. The sole purpose of this tandem position is that it allows the male to keep watch over the female and make sure she doesn't mate with any rivals until she has laid (what he hopes are) their jointly fertilized eggs on a suitable aquatic plant.

These highly demanding competitive conditions make it vital to keep your equipment in order. And the tiny fruit fly *Drosophila bifurca* is one insect whose equipment is beyond reproach. This little critter, a close relative of the same fruit flies that drive you crazy in your kitchen, is the proud holder of the record for the world's longest sperm: at almost 2.5 inches long, it is twenty times as long as the creature itself. For a man, that would be equivalent to having sperm the length of a handball court. How is it even possible?

The answer is that the whole thing consists largely of a thin tail bundled up into a ball. Enlarged photographs of the sperm look a bit like what happens when the kids make dinner and forget to put enough water into the spaghetti pot. So what's the point? Well, the long sperm are the fruit fly reproductive system's answer to Usain Bolt: the longest sperm outcompete the shorter and are more likely to win the race to fertilize an egg.

And since we're in the realm of the bizarre, there's no getting away from bedbugs—those bloodsucking rascals that hide out in cracks in the wall and beds in apartments and hotels the world over. When darkness falls, out they shuffle to poke their sucking snouts into you as you sleep. They're definitely not the kind of souvenir you want to bring back from your vacation, but the fact is that they are a growing problem all over the world. In part this is because we travel a lot, but the main reason is that bedbugs have developed resistance to the most common insecticides.

At any rate, the point is that the males of certain true bug species, including bedbugs, skip anything resembling foreplay; they can't even be bothered to find the female's genital opening, instead simply sticking their sex organ into her belly and leaving the sperm to find their way through the orifice to the egg cells. This often injures the female, preventing her from being able to mate with any other partners. In this way, the male attempts to ensure that he will be the father of her children. That said, the female has evolved a reinforced area on her belly where the male most often punctures her, which limits her injuries.

This illustrates an important point: the battle of the sexes involves two warring parties, and both sexes fight for what is most advantageous to them from an evolutionary point of view.

Ladies' Choice

It's possible that early insect researchers, who were almost exclusively men, had a tendency to see everything from the male perspective. Be that as it may, the fact is that modern research is yielding ever more examples of how female insects also work to advance their own interests.

One such example is the way certain females gobble up the male once they've finished mating. This is most common among spiders, distant relatives of insects. The male American fish spider, for example, dies in the middle of the act. This is because his sex organ bursts once he's delivered his sperm (or, to put it in dry scientific terms, "We observed that mating results in obligate male death and genital mutilation"). And then he is eaten—for the sake of the children. Even though his chosen one is a chubster fourteen times his weight, his little body still provides a useful protein boost. A little extra food comes in handy when you're preparing to lay hundreds of spider eggs.

The praying mantis is also famed for what is known as sexual cannibalism. That said, field studies have demonstrated that the male ends up on the supper menu more rarely when mantises mate in natural surroundings than in artificial laboratory conditions.

However, the insect mom has plenty of other tricks up her sleeve: it turns out that she can secretly control which males get to father her children. A great many mechanisms come into play here; the sperm's quest to reach the egg is more like an obstacle course than a gentle dip in calm waters. Since it is common for the sperm to be stored in a special "sperm bank" inside the female for the actual fertilization of her eggs to happen at a later stage, she has several ways of influencing which sperm to save and use.

One scientist carried out a cunning if brutal experiment to demonstrate this. She split a pile of flour beetles into two groups. Half the males were put on a starvation diet to make them look like weak specimens with poor genetic quality. Of the females, the scientist simply killed half so that they couldn't influence the outcome. When the scientist put the beetles together, both the starved and the well-fed males mated with living and newly defunct females, of which there were equal numbers. And now the really clever bit: the scientist found that the internal sperm banks of the dead females contained just as much sperm from the hungry, poor-quality males as from the well-fed, high-quality males. Those of the living females, however, contained a great deal more sperm from the high-quality males. This indicates that the females were taking active measures to control the process and ensure that the strong, high-quality males would father their children.

Life without Men?

There are plenty of ways to solve the conundrum of ensuring that "generation shall follow generation," and insects provide several examples of most options. Sexual reproduction, which requires both a male and a female, is the most common—among insects, too. But many insects can opt for the single life and still continue their line.

In fact, several insects are periodic practitioners of virgin birth. Female aphids, for example, can use this method to achieve a quick, effective baby boom on your rosebush in the spring. They don't have time to hang around and wait for their eggs to hatch, so they simply give birth to living aphid babies from egg cells that develop into new individuals without being fertilized. And that's not all: in some aphid species, the females can be like

Russian dolls: they contain baby aphids that are themselves already carrying new female aphids!

No wonder your rosebush is teeming with life! Despite the absence of men, it's questionable whether we can accurately call this "the single life." Soon there won't be enough room for all of them on the same bush. Up until now, the ladies have been wingless, but now it's time to squeeze out a few winged females who can fly over to the neighboring bush and continue the mass production there.

When the days grow shorter, the temperatures fall, and autumn is on the way, another change is triggered. The aphid ladies switch to production of males *and* females. These then mate, and this time the female lays eggs—the only way that aphids can survive the winter. She places the eggs on a suitable perennial plant. When spring comes, the eggs will hatch new virgin-birthing females. And the game will be on again.

Why is there any need for men if, in a single season, a female aphid can be sole progenitor of more children, grandchildren, great-grandchildren, and so on than there are humans on the planet? Wouldn't it be more productive if all individuals were able to produce offspring rather than just half of them? (Not to mention all the time that would have been saved if there were no need to worry about dating . . .)

Biologists have long been interested in the question of why most animals and plants come in two sexes, and the discussion is still ongoing. One disadvantage of virgin births is that all the individuals are genetically identical, which leaves species less room to maneuver if environmental conditions happen to change. Consequently, sexual reproduction, which blends the genetic material of two individuals, is a good and necessary means of promoting genetic variation and weeding out harmful mutations. Another handy thing about having two sexes is that it allows species to rely on different strategies: one sex may have few but large and nutritious sex cells—egg cells—while the other may have many small mobile sex cells—sperm cells.

Long Live the Queen!

Aphids aren't the only insects that live in a thoroughly female-dominated society. It is extremely likely that every single ant, stinging wasp, and honeybee you've ever seen was female. With very few exceptions, at any rate.

Do you remember *Bee Movie*, the film about Barry, the male bee who gets bored of life as a factory worker in the beehive? Biologically speaking, it's all wrong. So, for that matter, is Shakespeare's *Henry V*, in which he describes how the many inhabitants of the beehive are overseen by a bee king. The worker bees in the beehive aren't males. Nor are they ruled by a king bee.

The ladies are the ones who decide and do all the important work in the world of the honeybees. All the worker bees are female, and their ruler is a queen. The males, the drones, live for only a short period in the autumn and have just one role: to mate with a new queen. The male bees don't even gather their own food but are fed by the female workers.

Now perhaps we can forgive Shakespeare, DreamWorks, and the other people who have gotten this so badly wrong, as the misconception is old and hard to put to rest. The ancient Greeks tried to find out about the lives of bees but just couldn't get it all to add up. After all, they knew that normal honeybees had stingers—and surely females couldn't come equipped with such a formidable weapon? And if the short-tempered stingers were females, the large, sluggish individuals that couldn't even be bothered to collect nectar must be males, and that just couldn't be the case, could it?

It wasn't until the end of the 1600s, with the invention of the first microscope, that it became possible to establish that, yes, the tireless, terrifying workers and their monarch were all females, and the idlers were males.

But it would be another two hundred years before people truly understood how bees come into the world, because nobody had ever seen a honeybee having sex. The prevailing theory in those days was that the male bees, the dozy drones, engaged in the whole process at a respectful distance, remotely fertilizing their queen with what was fancifully referred to as "sperm odor."

Only in the late 1700s was it discovered that queen bees who had been out for a flutter returned to the hive with a male sex organ attached to their own genital opening, the remnant of the lucky winner, selected from a swarm of drones that chased her. The queen often mates with several members of the swarm. She saves up all the sperm cells (as many as 100 million) in a special internal sperm bank and doles them out as needed over the rest of her life.

For the drone, however, mating is the last thing he will do with his life. The actual transmission of sperm is nothing short of an explosion—so powerful that the drone's sex organ splits open and is torn loose from his abdomen, and he dies shortly afterward—a bit like a miniature version of "comes in like a lion, goes out like a lamb."

It's so extreme that it even inspired tabloids to devote a few column inches to insects, accompanied by headlines like this one from the British newspaper *The Sun*: "Male Bees' Testicles EXPLODE When They Reach Orgasm."

Beyoncé Was Right

From the bees' queen to the woman today's teenagers call Queen B—pop diva Beyoncé Knowles. Insects got a PR boost from an unexpected quarter a few years ago when media platforms worldwide broke the news that a new species of horsefly had been discovered and named *Scaptia beyonceae*.

There are two reasons why the Beyoncé horsefly got its name: first, because it was originally collected in the year of Beyoncé's birth, although that wasn't confirmed until long afterward; second, and more important, because it had such a beautiful backside. The trimming of golden hairs on its rump reminded the scientists tasked with naming the species of the artist's rear end when encased in a tight, glittering diva dress. (I eagerly await the day when there are more of us women entomologists, so we can start naming insects for their broad, manly winged shoulders or their ripped abs . . .)

I'm not sure how flattered Beyoncé was, if indeed she was even aware of the whole business, considering that the horsefly concerned was from inland Australia. Although horseflies are flower visitors and contribute to pollination, they are known primarily as nuisances to people and domestic animals; it hurts when they take a chunk out of us, they stress out animals, and they can transmit disease.

At any rate, around the same time as all this was going on, Beyoncé had a major hit that asked the question: Who run the world? Perhaps you know the answer: Girls!

I don't think for a second that she had insects in mind when she sang that. But she might just as well have. Because if we count up all the male and female animals on the planet, insects are responsible for ensuring that there are more girls on Earth. If we ignore bacteria, hermaphrodites, and other organisms without any clear gender and look at the proportion of females among the remaining animals, some extremely abundant groups, such as insects, are notably dominated by females. The honeybee workers are female, all 83 billion of them. All ant workers are female, and there are enormous numbers of ants on planet Earth; although there is no agreement on an exact figure, the BBC believes it is safe to assume that ants are the most abundant type of insect on the planet. And other fairly abundant insect species, such as aphids, may be female dominated at certain times of the year (see page 37).

Could this female domination on land be offset by aquatic species? The sea contains small crustaceans, the aquatic counterparts of insects, which dominate in number terms—animals such as *Calanus finmarchicus* and other types of copepods. The sex distribution among these crustaceans is more equal, but in this group, too, scientists sometimes report an excess of females. Even among cattle and poultry, which have a large presence on the planet, bulls and roosters are normally outnumbered by their female counterparts, thanks to human culling. Okay, there are also some organisms that typically have an excess of males, including some flatworms and tortoises, but it is unlikely that this is enough to help correct the imbalance.

So it looks as if Queen B was right in a way. Calculated by the number of total individuals, "girls" actually are the ones who make the world go round, thanks to insects and the extreme dominance of females in the most successful species.

I Am Fatherless but Still Have a Grandfather

How can social insects such as honeybees, ants, and many wasp species form societies with such skewed sex distribution? Part of the secret lies in how the sex of the offspring is determined in these insects. For human beings and many other insects, sex chromosomes determine the whole business, but these kinds of insects don't have sex chromosomes.

Sex is determined by whether the egg is fertilized or not—and the queen is the one to decide that. She's the only one allowed to lay eggs. If she fertilizes the egg with the sperm she has saved up from the swarming, it becomes a female—a worker or a queen, depending on the nourishment it receives during the larval stage. If she lays an unfertilized egg, it becomes a male.

This system leads to some genetic peculiarities, especially if the queen has mated with only one male in her life—because in that case, the queen's daughters will be more closely related to their sisters than to any offspring they themselves might have! Put simply, this is because every one of the bee dad's sperm contains exactly the same genetic material, so all his daughters (he cannot have sons—remember, they develop only from unfertilized eggs) inherit identical genes from him.

This makes it more advantageous for the daughters to opt out of having children themselves and instead help feed more sisters, including new queens. This is simply because that strategy will enable them to pass on more of their own genetic material.

For a long time, people thought this provided a good explanation for the social insects' strange societies dominated by sterile workers. But we now know that honeybee queens usually mate with several males. And in termites, which are also social, sex is not determined by whether or not the egg is fertilized—so that explanation simply doesn't hold up. Impassioned debate continues over what other mechanisms might explain the phenomenon.

At any rate, this strange system means that some unexpected challenges would arise if a male bee tried to draw his family tree. After all, he has no father, since he was born from an unfertilized egg. Yet he does have a grandfather, a maternal grandfather to be precise.

Our human hobby of genealogy—involving my children, your children, and our children—is a piece of cake by comparison!

Parental Leave the Insect Way

Insect moms generally consider their job to be done once the egg has been laid. But there are exceptions. Some insects are genuinely nurturing, providing variants of both bottle feeds and diaper changes. And these discoveries are incredibly useful beyond providing material for your next dinner party anecdote. By studying strategies used by related species that do and don't engage in child care or by manipulating species and observing the impact on the offspring's survival, biologists have learned a great deal about ecology and evolution.

For example, there is a cockroach (*Diploptera punctata*) that gives birth to living young. This means that the eggs hatch inside her, so the nymphs must be fed some nutrients in order to grow up big and strong. Cockroaches don't have a warm, cozy uterus where their young are fed intravenously through an umbilical cord. Instead, the mother has special glands in her abdomen that excrete milk protein in liquid form. The nutritional content of this "milk" is supposedly like battle rations—an optimal blend of proteins, carbohydrates, and fat. Some claim it could become the new superfood for us humans, too. But since milking cockroaches is pretty time consuming, we'll probably have to opt for producing the milk synthetically.

Another of our less popular insects, the deerfly, has a similar life cycle. It is a parasite that sucks deer blood and swarms at the height of the mushroom-picking season. Although it rarely stings humans, it is annoying when masses of them land on you, shed their wings, and crawl around in your hair. But for elks, they are a genuine problem. One elk that was examined by the Norwegian Veterinary Institute in 2007 was found to be infested with 10,000 deerflies!

Deerfly eggs also hatch inside the mother, and the larvae are "breast-fed" through special glands inside the maternal body, while the mother sits snugly ensconced in the elk's fur. Her offspring are "born" in the form of a sort of pupal cocoon, which becomes hard and black as an ebony bead, then falls off the elk and onto the ground. There it lies until it hatches the following autumn, when the cycle starts again.

Other insects also nurse and nurture their young. We have already taken a look at the social insects, where all the many sisters are employed as nannies for their younger siblings. And their mom is far from lazy, either. A termite queen drops a new egg every third second throughout her entire life—so it's no wonder she needs help from their older siblings.

Earwigs, those elongated brown insects with pincers at the rear, are particularly affectionate mothers. Though they may not exactly change diapers, they do keep watch over their eggs, cleaning away fungus spores and washing them with a substance that is assumed to inhibit the growth of mold and fungus. When the young first hatch, they fetch food and feed the small nymphs. An experiment showed that earwig moms' tender care multiplied the eggs that hatched from 4 percent to 77 percent. Burying beetles are another example of caring parents (see page 122).

And it's not all about the mothers. In Scandinavia, we are proud of our progress on the gender equality front. But when it comes to the smallest among us, other countries are way ahead of us on this issue—maybe because there isn't a single representative of giant water bugs (belostomatids) in Norway. Also known as toe-biters and electric light bugs, this subfamily contains a rare example of fathers being the ones who take parental leave. In fact, they take on a whole brood from different mothers. After mating, the female lays her eggs in neat rows on the father's back, and it's his job to look after them, floating on the surface of the water and making sure that the eggs neither dry out nor drown. And the mother? Like Henrik Ibsen's Nora, she goes her own way.

Some insects will go to extreme and savage lengths to raise their young, not by sticking around to look after them but rather by making sure there is fresh meat waiting when the larva hatches—by placing the egg inside another living being. In the next chapter, we will have a look at some of the strange ways insects eat—and are eaten.

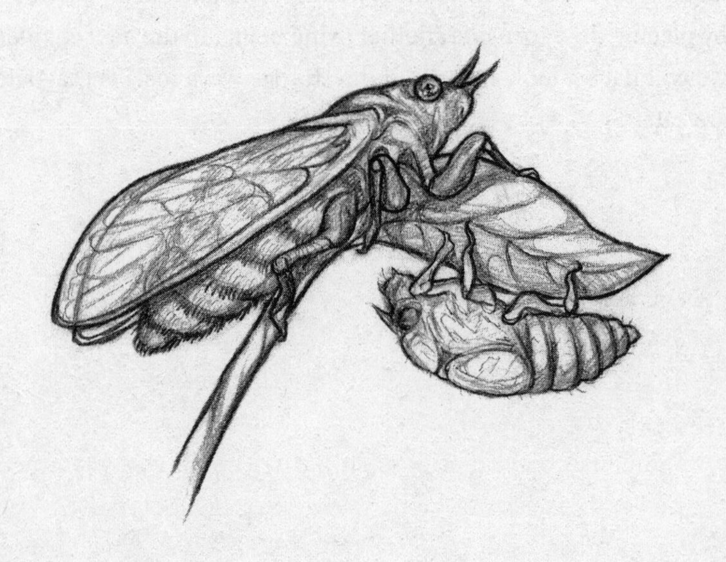

3

Eat or Be Eaten

Insects in the Food Chain

The recipe for a successful insect life is simple: you just have to stay alive long enough to reproduce. And to stay alive, you need food. Much of an insect's life consists of eating and trying not to be eaten.

Many insects eat one another. If there are fifty ways to leave your lover, I can assure you that there are an awful lot more ways of eating other creatures—including your lover. You can eat them from the inside or the outside. You can eat them as eggs, larvae, or adults. You can eat them using mandibles, sponges, or drinking straws. On the other hand, you can simply stop eating: quite a few insects eat only as larvae but don't feed at all as adults.

Since the objective is to keep on the right side of life's brutal but simple eat-or-be-eaten rule, insects go to extremes to avoid being gobbled up by other creatures. They may live in hiding, concealing themselves through camouflage or by pretending to be something else—preferably something dangerous or inedible. They can opt to survive by disappearing in the crowd or by collaborating with others in ingenious ways. Insects' strategies for laying their hands on nutrition without themselves becoming food is an object lesson in jaw-dropping but often brutal adaptations, which would be criminal of me to keep from you.

Darwin's Qualms

Take parasites, for example. Many insects are what we call parasitoids—parasites that ultimately kill their host. The host is often devoured from the inside out: the parasitoid larvae hatch inside an animal, for example another insect, and slowly but surely eat their way through all its internal organs. The whole thing is elegantly done: the larvae leave the vital organs until last. Well, we all prefer fresh meat, after all! The host usually dies once the parasitoid larvae have eaten their fill and are ready for adult life.

The natural historians and theologians of the 1800s tore their hair out when they found out about this. It just didn't fit in with their notion of a creation formed by a good and loving God. Darwin also struggled with it, writing to his American colleague Asa Grey in 1860, "I cannot persuade myself that a beneficent and omnipotent God would have designedly created the Ichneumonidae with the express intention of their feeding within the living bodies of caterpillars."

If only he had known! There are *much* worse things than that.

Zombies and Soul Suckers

The beautiful, green-eyed *Dinocampus coccinellae* is a parasite wasp. The female sticks her egg-laying tube, or ovipositor, into a ladybug and lays an egg. The egg hatches, and over the next twenty days the wasp larva chews its way through many of the ladybug's inner organs. Then the larva somehow squeezes its way out of the ladybug's abdomen while its unfortunate host is still alive. The wasp larva spins itself a little ball of silk between the ladybug's legs, where it transforms into a pupa.

Something quite remarkable happens next: the ladybug's behavior abruptly alters. It stops moving and just stands there, stock still, like a living shield. But every time a hungry foe of wasps approaches, the ladybug gives a jerk, thereby scaring off anything that might consider eating the now helpless monster that has just eaten it up. This lasts for a week until the wasp hatches and flies off, leaving the ladybug to its own devices.

The big question here is how the wasp mother can control the ladybug, transforming her into a zombie babysitter. After all, several weeks have passed since she laid her egg and vanished. The answer is that the wasp mother injects the ladybug not just with the egg but also with a virus. The virus accumulates in the brain and is controlled by a timing mechanism that paralyzes the ladybug at the precise moment when the larva is squeezing its way out. So the virus enables the wasp to take over the brain of the ladybug, making it serve not just as baby food but also as a babysitter. The only good thing we can say about all this is that, unbelievably enough, the ladybug sometimes survives the whole ordeal.

The cockroaches that fall prey to the soul-sucker wasp aren't so lucky. Do you remember the dementors in *Harry Potter*, those flapping black monsters that suck out people's souls? That's what gave the *Ampulex dementor* wasp its name. It is one of several species in the cockroach wasp genus, a representative of which can even be found in Norway. During childhood, these wasps live inside cockroaches.

Here, too, the whole process starts with a mother who's out and about wielding her egg-laying stinger. First she stings the cockroach in its chest to paralyze its legs for a few minutes—because the next stage involves high-level brain surgery, which requires "the patient" to lie completely still. Now the wasp stings the head. With extreme precision, she places a dose of nerve poison into two specific points of the cockroach's brain. This blocks the signals that control its ability to start moving: the cockroach can still move but cannot itself take the initiative to set in motion. The cockroach is now at the mercy of the wasp's will.

And the wasp's will is to take the cockroach to a place where she can lay eggs on it. But since the cockroach is far too big for the wasp to carry, it's handy that it has now lost whatever cockroach free will it might once have possessed but is still able to walk. This allows the soul-sucking wasp to simply bite into the cockroach antennae and lead her prey wherever she wants, like a dog on a leash—straight to death.

The cockroach is a biddable prey that allows itself to be led down into a hole in the ground; here the wasp lays an egg, which it glues to the cockroach's leg. Then the soul-sucker wasp blocks the entrance to the hole with small stones and vanishes. Her little larva child spends the next month fattening itself up. First it sucks the bodily fluids out of the cockroach's leg, and then it bores into the creature's insides and gobbles up its intestines, before forming a pupa inside the cockroach, which eventually dies.

Ugh! Maybe Darwin was better off not knowing about this. It's difficult to see any goodness in such ruthless behavior. Then again, evolution has never been driven by love and compassion.

Bold Hitchhikers

Some insects live off the young of other insects. The bold blister beetle eats bee larvae yet is still able to hitch a ride with the parents all the way into the nursery.

One May day, as I was sitting out in the sunshine, a strange fat beetle came puttering across my garden table, shimmering and bluish black. It looked as if it had borrowed a tailcoat three sizes too small for it: its abdomen was so full of eggs that it bulged out beyond the hind edge of the wings. It was a blister beetle paying a morning visit. In Norwegian, it's also known as a spring beetle, May beetle, or Easter beetle, and it is as rich in adaptations as it is in names.

The plump lady beetle is the source of what must be the spring's strangest stowaway. In a little while, she'll dig a hole in the earth and squeeze out a load of eggs, maybe as many as 40,000. The eggs hatch into larvae with hefty hooks on all six legs. They look a bit like long, narrow head lice or wingless stone flies and are filled with feverish energy. The triungulin larvae, as they are known, eventually gather in flowers, and there they wait for life's great lottery draw.

The thing is, it is crucial for these larvae to end up in the right place if they are to have a shot at life. And they need a ride to get there. They hook onto the first insect that lands in their flower—but it's game over for all the larvae that hitch a ride with the wrong kind of bee. And this is precisely why so many eggs are needed in the first place: the only ones whose future is ensured are the fortunate few who get lucky and stow away on a wild bee that's heading in the right direction.

Blister beetle triungulins gather in a flower in such a way that they form a shape that resembles a bee. The larvae also emit scent signals that imitate the smell of a lonely female bee. A male bee soon comes a-wooing. As he tries to mate with what he thinks is a female bee, she disintegrates and the triungulin larvae climb up onto the male bee. When the bewildered bee then flies on and, with luck, meets a real lady bee, the larvae hop onto *her* like rats leaving a sinking ship. That way, they are ensured a lift home to her hive.

The triungulins repay their driver by shape-shifting into legless larvae. They lie still in the hive, slurping up all their driver's pollen. And for dessert, they generally gobble up the wild bee larvae that are the rightful inhabitants of the hive. Once the blister beetle larvae are good and full, they pupate and wait for spring. That way, the cycle can start all over again.

Blister beetles owe their name to the fact that they secrete a blistering agent called cantharidin—one of the more poisonous substances known. An amount the weight of a grain of rice is enough to kill a human being.

For some reason or another, somebody got the (mistaken) idea that cantharidin was an aphrodisiac. Dried blister beetles of the "Spanish fly" species (*Lytta vesicatoria*), which are found further south in Europe as well as in the east, were once used as a sexual stimulant for men. It is said that Livia, the scheming wife of Emperor Augustus (of Nativity story fame), sprinkled her male guests' food with crushed Spanish fly in the hope that it would make them cast aside all discretion and self-control and do things she could later use against them.

In reality, the substance causes blisters and festering sores if it comes into contact with your skin, as well as painful irritation and swelling of the urethra if you eat it. What's more, there's an incredibly thin line between nonlethal and lethal effect. You don't want to mess with this stuff.

Blister beetles are adapted to emerge at the same time as the first flight of the solitary wild bees they parasitize. That's why you see them only in early spring. You're best off leaving a blister beetle to live out its peculiar life in peace if you happen to spy one.

Insects That Sing for Their Supper

I'm not much good at cooking Sunday dinner. We've often been on a hike, and nobody feels much like cooking when we eventually get home. And we don't know what to make, either; we were in no state to think two whole days ahead when we did the shopping on Friday afternoon, exhausted after a hectic week.

Oh, to be an insect at a time like this! Or, more precisely, a spotted predatory katydid, a big, bright green Australian bush cricket. It would sort things out in a jiffy *and* make sure the food was delivered to the door, good and fresh. So fresh, in fact, that it would deliver itself.

All katydids need to do is sing for their supper. It comes running, straight into the jaws of a poor soul starving for Sunday dinner. What do they sing? Well, put it this way—it's along the lines of Romeo's serenade beneath the balcony. The katydid has learned to imitate the mating signal of a totally different species, a classy cicada cutie, and this brings unsuspecting male cicadas strolling over. They head toward the sound, but instead of finding a sweet fellow cicada, they encounter a hungry and much larger enemy. Sunday dinner has just served itself.

In scientific language, this is called "aggressive mimicry"—a process whereby a predator or a parasite imitates the signal of another species in order to exploit the recipient of the signal. There are several examples: the *Photuris versicolor* firefly, for instance, can imitate a total of eleven relatives and pass itself off as a hot-to-trot female of all those species. Thus it is able to sit around idly, flashing away like a short-circuiting Christmas tree, and make its food come to it.

Even stranger is the bolas spiders' home delivery system. These spiders spin a thread with a sticky lump at one end, which they swing round and round until it hits a passing moth. The moth is then hauled in like a fish on a hook and packed up neatly in silk to be digested in peace and quiet when the night is over. This hunting weapon is reminiscent of a bolas, the implement consisting of two heavy balls linked by a cord that is used by gauchos, Argentina's answer to North American cowboys.

It's one thing to be a gaucho on horseback throwing your bolas at an animal you're chasing down; it's quite another to be a spider, sitting stock still. What are the actual chances of an owlet moth passing the spot where you are sitting still, seemingly innocent, swinging your bolas? Roughly equal to zero.

That's why this spider has also found a way to sing for its supper. It sings with scent. It has learned to imitate the complex scent signals of various species of moths. Sensing that love is in the air, Mr. Moth flies closer and closer to the source of this natural version of a femme fatale's scent—until it finds itself stuck fast in the spider's trapping gear.

Robber Flies Deserve Their Own Day

There is a day for everything: we have the World Migratory Bird Day, the World Day of Happiness, we even have Waffle Day and International Tea Day. But perhaps you didn't know that the last day of April every year is World Robber Fly Day. The creator of the #WorldRobberFlyDay hashtag, Erica McAlister, is an insect specialist at the Natural History Museum in London. She thinks that we ought to celebrate insects a lot more than we do. And why not start with robber flies?

Robber flies (Asilidae family) are pretty hefty predators. The family includes species that are up to 2.5 inches in length, which is nothing short of gigantic by fly standards. They are sun-loving, dark, often slender flies with powerful legs, enormous eyes, and a bushy mustache on their upper lip. They have full command of the air and can switch direction as they hover around waiting for unsuspecting prey to fly peacefully by.

In the blink of an eye the prey is trapped by six powerful, hairy robber fly legs. Without bothering to land, the robber fly thrusts a solid proboscis into its prey, which may well be an insect larger than itself—in warmer areas, it may even be a hummingbird. The robber fly injects its victim with a cocktail of spit, poison, and digestive fluids, instantly transforming the innards of the trapped creature into a kind of insect smoothie served up in a handy beaker. A couple of swift slurps—generally at top speed—and the robber fly casts aside the empty shell. Not for nothing are these roughnecks also known as "assassin flies."

Many robber fly species are rare, and we know little about how they spend their larval phase. As we know, they are important in controlling other insect populations and keeping them down, so it might be wise to learn more about these hefty flying predators and their role in the food web.

Swarmageddon

Imagine a red-eyed army of insects crawling up out of the earth, slow and silent. Every insect is the size of your thumb, and they emerge in such vast quantities that they call to mind a bad horror film about the end of the world. We're talking about a density of around 3 million insects on an area the size of a soccer pitch—but this is neither science fiction nor a doomsday prophecy. It is simply Swarmageddon, as some wits have dubbed the cyclical emergence of North America's seventeen-year cicadas.

These sap-sucking insects are quite content to forgo the outdoor life for as many as sixteen years in a row. They hide deep down in the dark alleyways and alcoves beneath the earth's surface, where they lie silently waiting. Now and then they'll take a sip of a root sap cocktail through the built-in drinking straw that serves as their mouth. Then, in year seventeen, the troops gather, preparing for some serious gate-crashing.

They emerge from the earth in hordes: pale brown, quiet, and wingless. The silent assembly climbs up into the trees and embarks on its final molting, which transforms the cicadas into adult individuals equipped for procreation. Voilà—out of the old exoskeleton steps a winged being, all dressed up and ready to party. The pickup sessions get under way, love is in the air, and silence is a thing of the past. If you've spent seventeen years lying quietly in the earth, you've got plenty to say for yourself. We humans hear the cicadas' song as an intense, high-frequency, grating racket. Multiply that by millions of singing male cicadas, and it's no wonder people can suffer hearing loss if they spend too much time out and about when the seventeen-year cicadas strike. The sound level can be as high as 100 decibels. Although the seventeen-year cicada doesn't sting or bite, people have to cancel garden parties and open-air wedding ceremonies when Swarmageddon strikes, because it simply isn't possible for people to talk together outdoors while this is going on.

Still, the party is short lived. After spending 99 percent of their lives underground for seventeen years, the cicadas' adult life is over in three to four weeks. Their song leads to mating, and mating creates new cicada eggs. The eggs hatch over the course of a few weeks, and small cicada nymphs crawl and crawl along the branch they were born on until it runs out and—boom! The newly hatched, wingless nymphs fall to earth and dig their way down—down into seventeen years of darkness.

Long before the nymphs hatch, their mom and dad are dead, having fulfilled their role. Now the only thing left to do is for people to fetch their snow shovels, clear pounds of lifeless insect carcasses off their drives and verandas, and await the next appearance in seventeen years' time—with anticipation or dread.

The seventeen-year cicadas are, in fact, the longest-lived insects we know of, along with their cousins the thirteen-year cicadas. There are several species, each of which may have several broods with different timing in different parts of the United States. No wonder the genus name of these peculiar insects is *Magicicada*.

Counting to Seventeen

So what is the point of the seventeen-year cicada's astonishing life story? And how in the world do the insects manage to count?

It appears that this behavior evolved because it reduces the chances of being eaten. Since cicadas are large and rich in protein, they are much-sought-after food for birds, small mammals, and lizards. This dramatic flooding of the food market ensures that a larger proportion of the cicadas are able to survive, mate, and lay eggs. It is quite simply a question of surviving by disappearing in the crowd. Since the time interval is so long, it is hardly likely that any predator will be able to adapt to it. And it is far from random that 13 and 17 are both prime numbers (numbers that can be divided only by themselves and 1). This means that a predator with a shorter cycle will never be "in sync" with the cicada boom cycle. Having a cycle that involves a pretty large prime number therefore reduces the chances of being eaten. This really is a fairly impressive math trick from an insect with the arithmetical ability of a toaster.

But how does the seventeen-year cicada know when it's time to put down its long drink and prepare to join the party on the surface? The trigger for their synchronized appearance is soil temperature. When the soil at depths of 8 to 12 inches remains above 65 degrees Fahrenheit for four days for the seventeenth time, the cicadas' internal alarm clocks all go off simultaneously. But we don't know quite how the countdown to year seventeen happens. Part of the explanation appears to be a biological clock in which chemical compounds alter over time. Perhaps external signals from the tree also play a role, with the cicadas "counting" the number of times it blossoms. Scientists who manipulated trees to blossom twice in one year did, in fact, find that seventeen-year cicadas emerged a year early.

There are singing cicadas in Europe, too, but they are not cyclical. Many people confuse cicadas (which are true bugs, order Hemiptera) with crickets and other grasshopperlike insects (which belong to the order Orthoptera; see page 27). Many of these also make a noise, but in different ways and at different times. The surging, intense insect sound you hear in the middle of a hot sunny day in southern Europe is typically the cicada.

Have you ever noticed little "spitballs" in the grass in summer? In many places, those spots of foam in the grass are known as "cuckoo spit," although they have nothing to do with birds. Inside the protective foam lies a meadow spittlebug, a distant cousin of the fat seventeen-year cicadas. European spittlebugs, which do not sing, spend the whole of their childhood at a foam party. The foam is created when the cicada nymph blows air through some slime it excretes from its rectum. This protects it against both predators and dehydration.

Why Do Zebras Have Stripes?

We can blame or praise insects for many things, and zebras' stripes may be among them—because in the same way insects have evolved to deceive predators and trick victims, larger animals have evolved in response to irritating insects. Actually, the mystery of these stripes has plagued biologists since Darwin's time. Why in the world are these particular animals striped when the same types of animal elsewhere are not? Scores of creative theories have been proposed over the years. Might the stripes provide camouflage for the animals when they are standing among small, scattered trees that cast a shadow? Maybe the pattern confuses predators, so they can't quite see where one zebra ends and the next one begins? Might the stripes have a cooling effect because the air warms up more rapidly over the black than the white sections, thereby creating tiny eddies of air? Or might the stripes serve a similar purpose as a conference name badge, letting the zebras know who's who?

This stripy debate hasn't yet been resolved, but some recent research rejects all these suggestions in favor of a fifth theory: the stripes repel insects.

Many infection-bearing insects live in the zebra's habitat, including tsetse flies and other kinds of biting flies that transmit diseases to large mammals. But if you are stripy, you get off lightly. Infection bearers don't like landing on striped surfaces. Why? Because stripes apparently confuse the insects' visual orientation, especially when the zebras are moving. The stripes create a kind of optical illusion, like the way we humans perceive the rotation of a spoked wheel or propeller as different from its actual rotation. So the new theory is that evolution has promoted the zebra's stripes because they lead to less insect trouble and consequently improved survival rates.

By the way, have you ever wondered what color the zebra is beneath its stripes? Well, its skin is not striped; it is black. In other words, the zebra is black with white stripes and not vice versa. There's a handy piece of trivia for your next cocktail party.

Insects as Guardians of Law and Order

Insects are staple food for birds, fish, and many mammals. At the same time, we know that insects often also eat one another, and this is absolutely crucial for keeping down the populations of what we think of as troublesome pests.

We know that an agricultural landscape where the fields are interspersed with varied flora provides a habitat for many of the pests' natural enemies. Similarly, woodland consisting of natural forest contains more predatory insects and parasites that keep spruce bark beetles and other pests in check than do managed forests. Predatory insects and parasites control the number of other small creatures in the forest. Swedish studies have found that the great spruce bark beetle, a species that can cause major damage to timber, has a great many more enemies in a natural forest with a variety of dead wood than in intensively managed forests.

In the garden, too, insects help keep things in order. Take the wasp, for example. A growing wasp's nest requires a lot of nourishment. It is said that a wasp can eat and eliminate two pounds of other insects from a garden measuring a couple thousand square feet—although the source of this claim is uncertain.

When it comes to spiders, however, we have fresh estimates of how much insect meat the massed spiders of the world actually devour in a year. And it is far from trifling: the planet's eight-legged insect population gobbles down between 400 billion and 800 billion tons of insects a year. That's more than the entire human population manages to polish off even if we combine meat and fish consumption.

To put it another way, the spiders of the planet could eat up every human being on Earth in a single year and still have room for more. Fortunately for us, they prefer to feast on the earth's many insects instead.

4

Insects and Plants

A Never-Ending Race

Although many insects are predators or parasites, most of them eat a plant diet in the form of either salad (living plants) or compost (dead plants, about which more in chapter 6).

There are many nuances to the salad diet: insects may eat nectar and pollen, seeds, or the plant itself. There may also be some advantages in this for the plant, such as pollination or the dispersal of seeds. Over 120 million years, insects and plants developed tightly in tandem. They are often mutually dependent, but at the same time it is a never-ending race in which each party is out to secure more advantages for its own side. This love-hate relationship has led to some peculiar forms of coexistence.

Drinking Crocodile Tears

The life of a herbivorous insect is no bed of roses. Plant tissue in general is pretty meager fare, low on vital substances such as nitrogen and sodium. For instance, while the dry weight of most insects is at least 10 percent nitrogen (sometimes much more), plants overall contain only around 2 to 4 percent nitrogen. This has a number of consequences for herbivorous insects. Many have a lengthy larval period to ensure that they can acquire enough nourishment before metamorphosing and facing the adult world. Other larvae (with a shorter larval period) concentrate on the most nutritious parts of plants, such as the roots (where some plants have tenant bacteria that capture nitrogen for them) or the flowers and seeds. By the way, this is exactly what we humans do too; think of our staple food such as cereals and legumes.

Many aphidlike insects, which subsist by supping nitrogen-poor plant sap, have to guzzle up massive amounts—relative to their small size—to get enough nutrition. This leads to a tremendous surplus of water and sugar, which they excrete in the form of what we often call honeydew—much to the delight of other creatures (see pages 74 and 79).

Plants also contain very little sodium, a substance that is crucial to the functioning of all animals' muscles and nerve systems, among other things. Though members of the deer family, all herbivores, can acquire sodium by licking salt stones set out for them by friendly humans, insects must find natural sources that are rich in sodium. This is why you often see colorful butterflies sitting around a puddle, slurping up mineral-rich mud as a supplement to their nectar-based diet.

And if you can't find a puddle, how about crocodile tears? In 2013, fascinated field biologists on a river excursion in the jungles of Costa Rica were able to film and photograph a beautiful orange butterfly and a bee drinking the tears of a caiman—a member of the crocodile family—each sipping from one of its eyes. It turns out that this method of acquiring vital salts from reptile tears is more widespread than we had thought; it is just rarely witnessed. Drinking crocodile tears is undoubtedly more colorful than slurping up puddles.

Willow: The Most Important Meal of the Spring

Pollination is a win-win activity that binds insects and plants together. The insect gets food in the form of sweet nectar or protein-rich pollen. The plants have their pollen moved from one flower to another, enabling fertilization and the development of new seeds. Although some plants rely on the wind for cross-pollination or are self-pollinating, as many as eight out of ten wild plants benefit from insects' visits to their flowers.

Some plants are especially significant "insect restaurants" because they supply nectar at a critical moment. The willow tree is one example. Normally, it lives a pretty anonymous existence in forest and farmland. But in the spring, it enjoys its fifteen minutes of fame—because this is when the bumblebee queen comes tumbling out of her underground bedroom, where she has lain like Sleeping Beauty since the previous autumn. And she is hungry—after all, she hasn't eaten anything all winter. But there's no one around to prepare a delicious breakfast for her. Not yet. All the worker bumblebees called it a day when the autumn chill arrived, along with the previous year's queen, and now it's up to this queen to start a new community. If she is successful, both she and we will eventually end up with food on the table—because bumblebees, wild bees, and other insects are, as we know, crucial for the pollination of our food crops (about which more in chapter 5). First, though, Her Royal Highness must find something to eat. And this is where the willow tree comes in, serving as nature's starting motor.

The willow isn't being idle once the snow is melting on the hillsides. At a time when the other trees and plants have barely started to consider what to wear this year, the willow is already fully dressed—a bit lightly clad, admittedly, because the leaves won't arrive for some time. But the flowers are what matter here, during the first trysts of spring. The male and female flowers blossom on their separate trees: the male flowers are the familiar soft gray catkins, which eventually turn bright yellow thanks to their pollen-bearing anthers; the female flowers are more discreet but contain more nectar than their male counterparts do.

And this is the queen bumblebee's great stroke of luck: a fortifying breakfast that combines protein-rich pollen with a strengthening source of sugar nectar, all served up in the willow trees. This provides the energy that is urgently needed when you're about to single-handedly set up a whole new community of pollinators.

Once the bumblebee queen has eaten her fill, she will find a suitable nest site, either underground or aboveground depending on the species. Here she will gather a ball of pollen in which to lay her first batch of eggs, before covering the ball with wax. Later, the newly hatched larvae will eat their way through this pollen-stuffed nursery. In the meantime, the queen is not being idle; she also builds a wax honey pot and fills it with regurgitated nectar. In this way she makes sure she can feed herself while incubating her eggs. Bumblebee eggs need to be kept at around 86 degrees Fahrenheit to develop properly, and the queen broods her eggs just as birds do. The queen actually has a bare spot on her abdomen to help disseminate the heat from her body to the eggs. In this first period she must sometimes leave the nest for short foraging trips, but as the colony grows, the workers take over the gathering of pollen and nectar and the queen concentrates on laying eggs.

Later in summer the bumblebee queen stops producing female workers. Instead, she lays unfertilized eggs that develop into males, and the larvae from her fertilized eggs are now fed in a way that makes them develop into new queens. With autumn drawing close, the new queens and the males mate. For the old queen, the males, and the rest of the summer's colony, it's game over. Only the new, now mated queen survives and crawls into a cozy space underground, ready for a long sleep, until spring awakens and the cycle begins again.

The Globeflower: Nature's Bed-and-Breakfast

Coupledom can be complicated, and that applies to the pollinating relationship between insects and plants too. The pollination of the globeflower is a case in point. With its bright yellow, almost closed head, the globeflower is easy to spot in meadow and waterside locations but not so easy to access.

Only three or four insect species, all belonging to a family known as globeflower flies, are able to find their way into this tightly packed miniature sun of a flower. But they are amply rewarded: it turns out that the globeflower is a bit like a bed-and-breakfast: it offers the visitors a hearty meal!

The globeflower provides the very best it has to offer: its own seeds. I'm not sure whether they contain as much protein as bacon and eggs, but they must certainly taste pretty good to an exhausted fly. Strictly speaking, the adult flies aren't the ones helping themselves to the grub, either. They just lay eggs in the ovules inside the flower, and that's where the larvae grow up. In fact, the only place they are able to develop is inside a globeflower seed.

So how in the world do globeflowers organize things so that a steady stream of flies can move from flower to flower with pollen? It's a question of collaboration and an ingenious balance between flower and fly. Because these particular flies are the only ones that can pollinate the globeflower, without their visits there would be no globeflower babies—in other words, no seeds. No wonder the flowers offer the very best they have at their disposal.

Yet it is an incredibly fine balancing act. If the flies eat up all the seeds, there will be no more globeflowers, and over the long term that means there will no longer be any hosts offering board and lodging—and that in turn means no more new flies. So it is crucial for the flies to lay eggs in a suitable proportion of the seeds. Quite how the flies work that out remains an open question. But the fact is that it works.

An Innocent Pizza Herb? Far from it!

Oregano is another example of the complex interaction between plants and insects—because this green herb, much used in Italian cuisine, is party to a cunning intrigue involving powerful alliances, disguises, and forgery.

Picture an arid, sun-drenched hillside in northern Italy emanating a heady scent of oregano, thyme, and marjoram. One of the oregano plants feels a tickling sensation around its nether regions: a bunch of *Myrmica* ants has decided to set up their nest beside the plant's roots. Now and then they gobble up a few small roots as they go about their work. This is hardly beneficial for the plant, which responds to the ants' munching by increasing its production of carvacrol, a substance that defends it against insects. Most ants have zero tolerance for the insecticide, but this particular species has learned to cope with it and stands its ground down there among the roots. We humans prize this defensive substance: carvacrol is what gives oregano its intense, powerful herbal scent.

But the aromatic substance has several functions. In Italian flower meadows, it also serves as an SOS, a kind of shout in the language of scent directed at a totally different species. The recipient is a beautiful butterfly known as the large blue. It lays its eggs on the plant, and the larvae then spend a couple of weeks there, developing and simultaneously preparing a disguise that would be the envy of any undercover agent. We're not talking false mustaches and hair dye here, because the visual aspect isn't especially important for ants. Scent is, however, which is why the butterfly larva veils itself in an alluring cloak of ant scent, perfectly adapted to the aroma of the ants living beneath the flower.

Next comes a critical moment: the larva lets go of the plant and tumbles to the ground. A *Myrmica* ant comes by on its way home from the eternal round of food gathering. It finds the butterfly larva, is tricked by the scent into thinking that it's a larva from the ants' nest, and carefully carries the butterfly larva into the darkness of the nest, where it is adopted on the spot. Even though it differs from the ant children in both size and color, it is nursed, cared for, and fed on regurgitated food by adult worker ants, which tend to it as diligently as they care for the nest's own children.

But the butterfly larva, which needs to multiply its weight by several hundred percent before it's done, is not content with recycled sugar water. As soon as its adoptive mothers turn their backs, the greedy butterfly larva tucks into the nest's stock of ant larvae. It supplements its aromatic disguise by imitating the sounds made by the ant queen—a kind of clicking song. This convinces the workers that the butterfly larva is a high-ranking ant, so none of them intervene while it runs amok in the nursery.

By the end, the butterfly larva has more or less polished off the entire colony. Peace returns to the area around the oregano plant's roots, and the larva can pupate. If it isn't reared in the right kind of ants' nest, the butterfly has no chance of producing future generations.

Who'd have thought that so much drama lay behind that scattering of green herbs on your pizza?

Seeds Play a Crappy Trick on Dung Beetles

In the case of oregano, both the plant and the butterfly benefit from the collaboration, but sometimes one of the parties has the upper hand and "tricks" the other—as when the red-tailed robber bumblebee *Bombus wurflenii* can't be bothered to crawl past the stamen buried in northern wolfsbane flowers to get to the nectar. Instead, it takes a shortcut, simply biting through the flower head and helping itself to the good stuff without having done anything whatsoever to earn it—because that way, there won't be any pollination.

Other times, the plant draws the long straw, as in the case of the reedlike plant *Ceratocaryum argenteum*, which grows only in South Africa. Cleverly enough, it produces seeds that look like dung: big round dark brown lumps identical in appearance to the leavings of the local antelopes.

Just as some clothing chains preperfume the garments they sell, this plant ensures that its "sales products"—the seeds—have an attractive scent—a scent of dung. Because it is targeting a very particular group of customers.

Normally, it's crazy for seeds to have a strong smell, since that makes it easier for hungry little seed-eating creatures to find and eat them. The explanation of this mystery came as a surprise. It was discovered by a group of scientists from Cape Town University who were actually supposed to be researching whether small rodents ate the strange heavy seeds. They set out nearly two hundred *Ceratocaryum* seeds in a nature reserve in South Africa, sort of like free samples. And as in the human world the whole thing had to be documented photographically, motion-sensitive cameras were set up beside all the seeds.

It turned out that the seeds were removed not by rodents in search of food but by gullible dung beetles that fell hook, line, and sinker for the seed's advertising offensive. The beetles believed that the smelly balls were antelope dung of the type they bury and then lay their eggs in.

Incidentally, dung beetles perform an extremely important service to the ecosystem by burying genuine animal muck, as this prevents pastureland from drowning in dung and ensures that the nutrients return to the soil (see pages 124 and 131). In this case, though, the beetles were tricked. They trustingly trundled away the spherical dunglike seeds and buried them an inch or so beneath the surface. At least a quarter of the seeds were thus sown in a new location: job done.

And what did the dung beetles get for their labor? Nothing.

The scientists hid in the bushes and dug up the seeds as soon as the mother beetle had shuffled off. They found no sign of eggs and no trace of any attempt to eat the seed, either. Apparently, the beetles ultimately discovered that they had been tricked and gave the whole thing up as a bad job. If beetles were capable of blushing, perhaps we would have seen the beetle mother's cheeks burning red when her naïveté was revealed live on camera. Just imagine being taken in by a reed! A pretty crappy trick!

Seeds That Serve Up a Packed Lunch for Ants

There are plenty of other plants that get insects, mostly ants, to disperse their seeds for them in return for a reward. We know of this in more than 11,000 different plants, or almost 5 percent of all plant species. It is common for the plant to ensure that there is a sort of payment in the form of a valuable supplement: a packed lunch for the ant. The ant carries the whole thing home to the anthill, but when the packed lunch is being served up to the hungry ant babies, the seed is thrown away, generally beneath the earth in or near the anthill. Some of the seeds are also lost in transit.

Ants help many plants, including common cow wheat, violets, and wood anemones. One cunning adaptation may be for plants to flower and produce seeds early, before the ants have much else worth eating, as this increases the chance of getting help with transport. The next time you see a common hepatica in spring, take a closer look when the flowers fall off, and you'll see the little packed lunches sitting on each seed.

Other plants have taken the collaboration with ants a step further, not just serving up food but also building the ants a house.

Acacias are the classic example: some develop enlarged thorns where the ants can live and provide them with nutritious food in the form of small packets of oils and proteins. In return, the ants keep hungry herbivores at bay and graze on competing vegetation around the acacias.

Wood Wide Web—An Underground Internet for Plants

Collaboration may be the smartest option when insects are on the warpath. Here plants are helped by an entirely different species: fungi. There's a lot more to the chanterelle or porcini mushroom than the cap that catches your eye while you're out mushrooming in the autumn. A large part of these mushrooms is hidden beneath the forest soil, forming the wood's concealed communication system—a network of fungal threads that links trees and plants, allowing them to communicate. Yes, communicate. We are constantly learning more about this close collaboration between fungi and roots, known as *mycorrhiza* (literally "fungus roots"), which we find in 90 percent of all the plants on Earth.

One thing this relationship does is help plants grow, as the fungi transfers water and plant nutrition from the soil. We have known about that for a long time. But the fungal network can also be used to send messages—about an insect attack, for example. Just like a school sending emails to all parents when the school nurse finds head lice in sixth graders or the public health warnings about the onset of this year's influenza bug, a plant attacked by insects can send chemical signals via the subterranean internet to say "Watch out—here come the aphids again!"

In an ingeniously constructed study, British scientists planted beans and allowed some of the plants to develop fungus roots but stopped others from doing so. Next, they eliminated the possibility of sending signals through the air by wrapping the plants in special bags that prevented signaling molecules from getting through. The next step was to let aphids loose on certain of the plants. According to the scientists' findings, plants that didn't get chewed themselves but *did* have contact with the plants under attack via the fungal internet developed defensive substances to protect them against the aphid attack. The isolated plants did not.

In forests trees also use this underground internet—call it the Wood Wide Web if you will—to send one another carbon. Some scientists think that the oldest and largest trees in the forest, the "mother trees," help the young saplings in the early phase of their life by sending a kind of food parcel through the network. Even trees of different species may send one another nutrition in this way. Perhaps we need to reconsider the way we think about forests: individual trees may be more closely linked than we had realized.

Working Your Land

Agriculture and animal husbandry are the basis of our modern civilization. They have enabled high population density, with all the opportunities that entails. But we humans were pitifully late starters compared to insects. Our agricultural revolution happened only ten thousand years ago. By then, ants and termites had already been practicing agriculture for 50 million years, and ants had been keeping livestock for twice as long. Is it any wonder that ants beat us hands down when it comes to the number of individuals on the planet and that the combined weight of these minute but multitudinous six-legged creatures matches the weight of all the human beings on Earth?

Insects don't grow plants, they grow fungi—specially adapted fungi that grow only in the ants' fields, in the same way as our crop plants have adapted to a life "in captivity." In South and Central America, leaf-cutter ants are common. Long columns of workers march out to cut off suitably sized sections of leaves and bring them back to the nest in the earth. The machinery that then takes over is so well oiled that it would exceed the wildest dreams of any industrial magnate. A long column of ants, all of slightly different sizes, do exactly what is needed—without demanding longer lunch breaks, better shifts, or higher pay.

The leaves are chewed up and distributed across the "kitchen garden." Other smaller ants lick the mass of leaves, thereby transferring fungi from the established parts of the garden. Even smaller ants shuffle carefully around in the garden, removing "weeds," which in this case mean bacteria or the wrong kinds of fungus. Once the fungus has grown and spread across the new section of the garden, certain ants work to harvest the nutrient-rich parts of the fungus and send the candy floss–like food around to all the rest, including the growing generation of ant larvae.

As in a well-run factory, this sort of streamlined production requires good access to raw materials. In the course of a year, an average leaf-cutter ant colony clears and maintains 1.7 miles of ant paths, which radiate out from the colony like the spokes of a bicycle wheel.

The agriculture practiced by termites resembles that of the leaf-cutter ants, but in this case the nest is built of earth and wood pulp mixed with spit and lies partly beneath and partly above the ground. A sophisticated air-conditioning system ensures that the temperature is maintained at optimal levels in the subterranean fungus gardens (see page 155). And termites don't bring in green leaves: they carry home sticks, grass, and straw. With the aid of their fungus partner, the plant matter is broken down and converted into more digestible termite food. The two parties, termite and fungus, are dependent on each other.

Certain bark beetles that live in wood also rely on fungus. This enables them to convert cellulose into edible material. These ambrosia beetles, as they're called, pretty much take a boxed lunch with them when they move into a new dead tree: they have special cavities in their body (mycangia) where they store a certain type of fungus. Once installed in their new accommodations, a dying or newly dead tree, they aren't content just to lay eggs in a crack; no, they excavate splendid chambers and corridors beneath the bark, and there they plant the fungus that will be grown like a kind of kitchen garden to provide healthy, nourishing food for the beetle babies. And this is probably necessary, since beetles' family life is not quite like our own. Although some ambrosia beetles stay in the tunnels to take care of their offspring, other species leave the kids to butter their own bread, so it's a good thing they have at least taken the trouble to fill the pantry before leaving.

We don't know how ants and termites manage to maintain high, stable production, even in such an extreme monoculture where they cultivate only a single species. It would be good news for our own future food production if we could wheedle this secret out of the insects.

Aphids as Dairy Cattle

Ants' animal husbandry is no less impressive. As described earlier (see page 66), aphids produce large amounts of sweet liquid, and some ant species provide bodyguard services in return for this substance. For the ants, easy access to carbohydrates is so attractive that they happily, and aggressively, defend their herd of "sugar cows" against anything that might even think of eating them. An ant colony can easily harvest 22 to 33 pounds of sugar from aphids over a summer; some estimates go as high as 220 pounds of sugar per anthill per year.

Ants have also been found to "herd" their cattle by restricting the aphids' capacity to disperse to other plants. Just as we humans clip the wings of geese and other winged livestock, ants may bite the wings off aphids. They can also use signaling chemicals to prevent the development of winged individuals or limit how far the aphids stray on foot.

The ants' nurturing of these sap-sucking insects can be a disadvantage for the host plant—hardly surprising, perhaps, since aphids and their relatives vacuum up huge amounts of plant sap. Some US scientists found proof of this when they were actually supposed to be studying the codependent relationship between ants and tiny cicadalike insects called treehoppers on yellow rabbit brush shrub in Colorado. To their irritation, black bears kept turning up and destroying the ants' nests in some of the areas they planned to study (taking a chunk out of the field equipment along the way).

In the end, the scientists decided to shift their focus and look at how the bear influenced the system instead. They then discovered that the plants grew better where the bear was present, owing to a sophisticated domino effect. When the bear ate the ants, there were fewer ants to frighten off the ladybugs. That meant there were more ladybugs. Since these were now left to eat in peace, they helped themselves to the treehoppers and other herbivores. As a result, there were fewer bothersome insects on the plants, which therefore grew better. That is how the presence of bears can improve plants' growth—by keeping ants in check!

Small Creatures, Great Significance

Connections don't always work the way we think they do. One example of this stems from wheat fields in the arid parts of Australia. In this case scientists wanted to study the positive contributions of insects, especially ants and termites. So they compared the wheat harvest in fields where these insects had been eliminated by insecticide and fields where ants and termites were allowed to remain.

It turned out that the wheat harvest rose 36 percent where crops had *not* been sprayed. Why? In areas as arid as this, there are no earthworms, so ants and termites do the earthworms' job instead, creating corridors that allow more water to trickle down into the soil. The water content was twice as high in the soil where these insects were allowed to live as in the soil where they had been eliminated. In addition, the nitrogen content was much higher. This may be because termites' guts contain bacteria that capture nitrogen from the air.

And as if it weren't enough that the insects improved the soil's supply of water and nutrients, seed-eating ants also ensured that there were only half as many weeds in the unsprayed fields as in the sprayed fields.

We can find other examples of the significance of ants in Europe. A Swedish study of coniferous forests shows how tiny ants might affect massive things, like the climate, by influencing carbon storage in forests.

Take a walk in any woodland and find yourself an anthill. This is the home of wood ants—hill-building ants of the *Formica* genus. In an experiment conducted in northern Sweden, scientists excluded these ants from small areas of the forest floor. This had major consequences.

The entire plant community changed. The four most common herbs became even more common. This increased the supply of nutrients to the soil in the forest because woodland herbs such as cow wheat and *Linnaea borealis*, or twinflower, decompose more easily than woody berry shrubs. The increase in nutrients was like putting a rocket under the tiny caretakers of the woodland soil. Most notably, the activity of various bacteria increased. This also led to the decomposition of old and seasoned remnants of dead plants.

So what was the net result of keeping wood ants at bay? Well, because the changes in the decomposition community meant that old, stored carbon was now suddenly being broken down, the scientists observed an overall 15 percent *decline* in the forest soil's storage of carbon and nitrogen. If this result holds true when scaled up, it means that in the absence of ants, a substantial proportion of the large carbon stock in boreal forest soil would be lost. Bearing in mind that northern forests cover 11 percent of the earth's surface and store more carbon than any other types of woodland, it is clear that, despite their modest size, wood ants have a major influence on fundamental processes such as the circulation of nutrients and carbon storage.

A Troublesome Cactus

We humans have long exploited the close relationship between insects and plants, and between predatory and herbivorous insects. Ancient Chinese documents, apparently from around the year 300 BCE, tell farmers how to move the papery nests of a certain ant into lemon groves to reduce the pests on the lemons. It was also common to set up bamboo "suspension bridges" between the trees to make it easier for the ants to move from tree to tree and keep away the pests. This appears to be one of the first examples of what we call biological control: the use of living organisms in the battle against pests, as an alternative to using chemicals.

We have moved species from one end of the planet to another, often quite intentionally, and with extremely variable results. Sometimes, things have gone terribly wrong—as in Australia in the 1800s, when somebody had the bright idea of setting up production of cochineal dye (see page 147) and optimistically imported a few shiploads of prickly pears from Mexico. Cochineal production went down the drain, but the prickly pear spread like wildfire. By 1900, the cacti covered an area the size of Denmark. Just twenty years later, the area was six times as large. A region the size of Great Britain was totally unusable for grazing or crop growing because it was overgrown by spiky cacti. It was a crisis. The authorities offered a generous reward to anybody who could come up with a way of combating the prickly pears. The reward was never claimed.

In the end, one world war and much desperation later, the solution arrived—in the form of an insect from South America. A mothlike lepidopteran of the *Cactoblastis* genus, whose larvae chew passageways through prickly pears, was brought in, tested, and bred in massive numbers. A hundred men in seven trucks drove around the whole of Queensland and New South Wales handing out paper quills filled with *Cactoblastis* eggs to the landowners. In the five years from 1926 to 1931, more than 2 billion eggs were distributed.

It was a spectacular success. By 1932, the moth larvae had killed off the cacti in large parts of the fallow land. This is still one of the prime examples of successful biological control.

But there's always another side to the coin. After the success in Australia, the moth was used to biologically control cacti in several other places, including the Caribbean islands. From there, *Cactoblastis* moths have spread to Florida, where they now threaten to wipe out unique local cacti.

5

Busy Flies, Flavorsome Bugs

Insects and Our Food

So you say you don't like insects? Well, then, maybe you don't like chocolate, marzipan, apples, and strawberries, either. These and countless other foodstuffs can be produced in the quantities and quality to which we are accustomed only with the help of insects. What we are talking about here, of course, is insects' work in the field of pollination.

Insects' visits to flowers contribute to seed production in more than 80 percent of the world's wild plants, and insect pollination improves fruit or seed quality or quantity in a large proportion of our global food crops. Although wind-pollinated crops (such as rice, corn, and various other grains) account for the majority of our energy intake, insect-pollinated fruits, berries, and nuts are important energy boosters, as well as being a vital source of variety in our diet. We know that the species richness of wild pollinating insects matters, too: a study of forty different crops across the planet showed that visits from wild insects increased crop yields in all systems.

And we are cultivating a steadily increasing amount of crops that require pollination; according to the Intergovernmental

Science-Policy Platform on Biodiversity and Ecosystem Services (IPBES), the volume of such crops has tripled over the past fifty years—but the occurrence and species diversity of wild pollinating species are simultaneously declining.

Some pollination also results in a by-product, and one in particular that we all know and love is honey—a natural sweetener with a long history. And if you fancy supplementing your diet with a spot of environmentally friendly protein, why not eat the insects themselves? They're packed with nourishment and form part of a normal human diet in most areas of the world—except the West.

In this chapter, we'll look take a closer look at insects' role in our food supply.

Sweet Stuff Steeped in History

We love sweet stuff. The US average per capita sugar consumption hovers at around 100 pounds per year. This is hardly surprising because the difficulty we humans have resisting a bowlful of sweets is deeply rooted in us. Once upon a time our ape ancestors lumbered their shaggy way around Africa eating fruit. Since the sweetest, ripest fruit had the highest energy content, we gradually evolved a preference for sweetness. Back then, it simply made sense to have a sweet tooth.

Anybody who has ever accidentally left a banana in a gym bag knows that ripe fruit has an extremely short shelf life. But there is another source of sweetness that is much less perishable and has long been in use: honey. In 2003, construction work on Europe's second longest oil pipeline unearthed honey jars in the 5,500-year-old grave of a woman in the Republic of Georgia.

So what exactly is honey? It is created when bees extract nectar from flowers, collecting it in a honey sac, a special pouch that lies between the esophagus and the stomach. This prevents the nectar that will later become honey from mingling with the food passing through the bees' digestive system. Once inside the honey sac, the nectar mixes with the bees' enzymes. When the bees return to the hive, they regurgitate the contents of their sac, passing it on to other bees, who store it in their own honey sacs, transport it farther into the hive, and regurgitate it into the mouths of yet more bees. In the end, the honey is stored in wax cells, where it remains for later use—or until we humans harvest it.

Hallucinogenic Honey

In the Spider Caves (Cuevas de la Araña) in Valencia, Spain, 8,000-year-old cave paintings depict the harvesting of wild honey. They show a man dangling from a rope or vine surrounded by swarming bees, one hand holding a collecting basket and the other inside the nest.

In Asia, vestiges of bee- and honey-based cultures still persist in the food as well as the culture and economy. Twice a year the honey hunters in the foothills of the Himalayas harvest the honey of the Himalayan giant honeybee (*Apis dorsata laboriosa*), the world's largest honeybee. It is a hazardous venture that involves climbing up high cliffs using ladders and ropes amid the buzz of cantankerous bees. These days, pressure from tourists keen to witness the phenomenon is causing overharvesting of the bee colonies; at the same time, erosion and the shrinkage of wilderness areas are altering the surrounding landscape, all of which are having adverse consequences for the bees. Furthermore, it did little to diminish attention levels when journalists discovered that one variant of the honey gathered in the mountains of Nepal has hallucinogenic properties. The reason for this is that the bees gather poisonous nectar from plants such as rhododendrons and bog rosemary or related heather plants. As a result, the honey may contain a poison called grayanotoxin, which not only affects your pulse and makes you dizzy and nauseous but can also cause hallucinations.

"Mad honey" is a recognized phenomenon in Europe as well. Accounts from antiquity tell of a disastrous military campaign in around 400 BCE, during which thousands of Greek soldiers retreating through present-day Turkey helped themselves to some wild honey. Despite the absence of enemies, their camp soon looked like a battlefield. According to the ancient Greek military commander and writer Xenophon, the soldiers raved like drunkards, quite out of their wits. Diarrhea and vomiting rampaged through the camp, and only some days later were the soldiers fit enough to struggle to their feet and resume their homeward march.

Other sources from antiquity describe the use of hallucinogenic honey as a weapon of war. A few honeycombs of rhododendron honey are ever so casually placed in the enemy's path—for who can resist a spot of the sweet stuff when they happen to stumble across it? The intoxicated soldiers are then easy to pick off.

This kind of honey is still produced in parts of Turkey, where it is known as *deli bal*. But there's no need to worry about poisoning yourself by eating "mad honey." It's vanishingly rare for the concentration in the modern, commercially produced honey to be high enough to have ill effects. Fortunately.

Beyond that, honey has long been prized for its antibacterial properties. Historically, it was used on wounds, and it is said that when Alexander the Great died in Babylon, he was submerged in a honey-filled coffin to preserve his body over the two years it would take to transport him to his burial place in Alexandria. The truth of this story is difficult to establish.

The Sweet Taste of Teamwork

One story that is certainly true, however, despite sounding pretty incredible, is the tale of a bird called the greater honeyguide (with the apt Latin name *Indicator indicator*). This African species helps us humans to find honey. The honeyguide is fond of both honey and wax and wouldn't say no to a smattering of bee larvae, either, and it is famous for its unique behavior. As its name suggests, it can tell other animals and humans where to find the honey. In return, it counts on receiving a share in the booty once the nest is broken up by something larger and stronger than itself.

Whereas most birds fly away on our approach, the opposite happens with the honeyguide. It seeks out humans, twitters, and then flies a little way off to see if they're following. New research has shown that the birds respond to certain human sounds. The Yao people are a tribe in Mozambique who still find honey in collaboration with the honeyguide. When the scientists played back recordings of the Yao people's special calls, this increased both the probability that the honeyguide would appear and the probability of its leading them to the bee's nest. The overall probability of finding honey rose from 16 to 54 percent compared to searching without a honeyguide. This is one of very few examples of mutual, active collaboration between wild animals and humans.

We have known of this peculiar collaboration since the 1500s, but some anthropologists think it may go all the way back to the days of *Homo erectus*. In that case, we're talking 1.8 million years. That tells you a bit about how sought after this product of the insect world has been for both animals and human beings for thousands of years.

Manna, the Miracle Food

Insects have other sweet tidbits to offer, too. And they could well be the origin of manna, the miracle food mentioned in the Bible—unless we think of it as a purely miraculous product, that is. According to the Old Testament, manna was the food the Israelites survived on during their journey from Egypt to Israel. That was something of a trip: a forty-year expedition through the barren Sinai Desert with few opportunities to lay their hands on food.

That thought also struck the Israelites: "And the children of Israel said to them, 'Oh, that we had died by the hand of the Lord in the land of Egypt, when we sat by the pots of meat and when we ate bread to the full! For you have brought us out into this wilderness to kill this whole assembly with hunger.'"

But the Lord, who had already taken care to equip the world with "everything that creeps on the earth" in the Book of Genesis, had a solution: "Behold, I will rain bread from heaven for you." "And when the layer of dew lifted, there, on the surface of the wilderness, was a small round substance, as fine as frost on the ground. So when the children of Israel saw it, they said to one another, 'What is it?' For they did not know what it was. And Moses said to them, 'This is the bread which the Lord has given you to eat.'" "And the house of Israel called its name manna. And it was like white coriander seed, and the taste of it was like wafers made with honey." "And the children of Israel ate manna forty years, until they came to an inhabited land." (All quotations from Exodus 16, New King James version.)

A slightly monotonous diet, perhaps: forty years of nothing but honey wafers would be enough to cure even the sweetest tooth. However, it seems to have been satisfactory hiking food because the Israelites did reach their destination after all. But are there any natural edible products in this part of the world that might have inspired the description of manna, the miracle food?

Proposed answers, with varying degrees of probability, range from the sap of various bushes or trees, from the flowering ash (*Fraxinus ornus*) to hallucinogenic mushrooms (*Psilocybe cubensis*), fragments of lichen (*Lecanora esculenta*, or *manna lichen*), or algae (*Spirulina*) carried by the wind to mosquito larvae, tadpoles, or other aquatic animals whisked away by tornadoes.

The strongest hypothesis is that manna may have been crystallized honeydew from a sap-sucking insect, specifically the Tamarisk manna scale (*Trabutina mannipara*). This little insect belongs to the scale insect family and sucks sap from the tamarisk tree (*Tamarix* genus), which grows extensively across the Middle East.

Because the sap sucked up by the manna scale (and many other sap-sucking insects, see page 66) is much heavier on sugar than on nitrogen, the insects have to jettison the surplus sugar. They do this by excreting a sugary secretion called honeydew. Large quantities of this sweet substance can accumulate on the tamarisk trees and dry into sugar crystals. People in Iraq and other Arab countries still gather this sugar from tamarisk trees and consider it a delicacy.

If this is the origin of the biblical manna, we might imagine that the wind loosened the crystals and sprinkled them onto the earth, making it appear that sugar crystals were raining from the heavens.

Marathon Food

Perhaps the Israelites ought to have taken some hornet juice with them on their long and arduous journey, too. It has been shown that the larvae of an Asian hornet produce a substance that is nowadays marketed as a miracle product to boost sporting endurance and performance.

Adult hornets cannot eat solid protein. So instead, they fly home to the nest and feed their larvae small chunks of meat. The larvae have mouths with teeth and can chomp away. In exchange for the chunks of meat, the larvae regurgitate a kind of jelly that the adults then slurp up.

Once people grasped that the contents of this jelly were crucial for the adult hornet's endurance—they can fly 60 miles a day at a speed of 25 miles per hour—it wasn't long before a commercial product aimed at athletes came on the market. Does it work? Debatable. But it certainly sells—especially since Naoko Takahashi won the Olympic gold medal in the women's marathon in Sydney in 2000 and gave much of the credit for her victory to the hornet extract. Sales really took off then! Today you can buy sports drinks containing hornet larva extract in Japan and similar products in the United States.

Billions of Starving Locusts

Sometimes insects simply eat our food. Locust swarms have been and remain a feared example of just that. In the Bible, swarms of locusts are described as one of the ten plagues God inflicted upon Egypt.

"So Moses stretched out his rod over the land of Egypt, and

the Lord brought an east wind on the land all that day and all that night. When it was morning, the east wind brought the locusts. And the locusts went up over all the land of Egypt and rested on all the territory of Egypt. They were very severe; previously there had been no such locusts as they, nor shall there be such after them. For they covered the face of the whole earth, so that the land was darkened; and they ate every herb of the land and all the fruit of the trees which the hail had left." (Exodus 10:13–15, New King James version.)

One fascinating aspect of this biblical quotation is that it remains accurate to this day in strictly ecological terms. Only when the *khamsin*, a warm southeasterly wind, has blown for at least twenty-four hours can swarms of locusts reach Egypt from the areas where they originate further east.

And it is a truly terrible sight. A single locust can eat its own body weight in food every day. And once we grasp that a swarm can contain 10 billion of these starving, flying, jumping creatures, distributed over an area about the size of a city of a half-million people, we begin to understand how the sky would be darkened and why they left not a single stalk of grass behind them.

Locust swarms still appear at irregular intervals, mainly in Africa and the Middle East, although estimates indicate that swarms are capable of affecting up to 20 percent of the earth's land surface. The mechanism behind locust plagues is like a one-way version of Dr. Jekyll and Mr. Hyde. Normally, the locust is a shy creature that does not cause any harm to crops. But when special weather conditions make their numbers surge, space becomes tight and they repeatedly bump into one another, which triggers a hormone that changes both the way they look and the way they behave in a matter of hours. They grow bigger, darker, and hungrier, and all of a sudden they feel strongly attracted to one another. Large bands of restless locusts form, moving across the landscape and meeting up with other bands to form even bigger groups. One theory is that starvation can lead to cannibalism in locusts and that the swarming behavior has evolved as an alternative.

Yet the fact that insects eat plants we ourselves want to eat is not solely negative. Many food crops we prize for their acidic, bitter, or strong taste have developed these flavors as a defense against grazing, by insects among other creatures. Think of herbs such as oregano (see page 70), the peppermint tea you're so fond of, or the mustard you squeeze on your hot dog. If the plants no longer needed a defense, they would spare their resources for other purposes and the taste might change. Many of the active ingredients in medicines extracted from plants may have originated from the plants' need to avoid being eaten by insects and larger animals.

Chocolate's Tiny Best Friend

We humans love our chocolate. Worldwide chocolate consumption is steadily rising, and Americans polish off around 11 pounds per person of the stuff a year! At the same time, producers are now warning of a possible chocolate shortage in the near future, citing global causes such as climate change and increased chocolate consumption in countries such as China and India.

There's actually one teensy-weensy factor nobody is discussing that is crucial to ensuring your access to chocolate: a little biting midge that is, smaller than the head of a pin. It's friendless. But perhaps it isn't so easy to make friends when you're the size of a poppy seed and all your relatives are bloodsucking jerks. Because the midge in question belongs to the family of biting midges, those teeny insects, also known as "no-see-ums," that crawl through your mosquito net, find their way into your ears and behind your glasses, and just can't resist extracting a few drops of warm blood.

Despite all that, this tiny insect is almost single-handedly responsible for the coating on your cookies, the chocolate bar that sweetens your hike, and the cup of cocoa that warms your bones on a chilly winter's day. Because out in the rain forests, this relative of the biting midge—the chocolate midge—has forgone blood in favor of a life spent crawling in and out of cacao flowers.

These beautiful blossoms, which grow straight out of the stem of the cacao tree, are very intricately constructed. The chocolate midge is one of the few insects that can be bothered—and is small enough—to crawl into a cacao flower and get the pollination done.

But the romantic relationship between the cacao tree and the chocolate midge is complicated because it won't do to use pollen from another flower on the same tree; no, pollination requires the right stuff, from a neighboring tree. If you consider that our new insect friend can barely carry enough pollen to pollinate a single flower, that it is bad at flying, and that the flowers last only a day or two before falling off, you'll get an idea of just how tough this particular relationship is.

In addition, the chocolate midge has certain stipulations when it comes to the fittings in its home. It needs shade, high humidity, and a layer of rotten leaves on the floor of the nursery. This is because its larvae develop in humid compost on the floor of the rain forest.

So the process doesn't produce a great deal of cacao—even less when the growing is done in open plantations, which are too dry and too far from the shade for the midge's liking. Only 3 of 1,000 cacao flowers in plantations are successfully pollinated and go on to become mature fruit. On average, a cacao tree produces enough cacao beans for barely 11 pounds of chocolate in its entire twenty-five-year lifetime.

To convert that into a more recognizable currency, this means that three months' worth of the entire production of a single cacao tree is required to produce, say, a Kit-Kat—plus the hard pollinating labor of hordes of industrious chocolate midges.

Marzipan's Midwife

Marzipan is simple: take some finely ground almonds and confectioner's sugar, and mix in a smidgen of egg white to bind it all together. Yet marzipan owes its existence to a pretty complicated "birth" that takes place in sunny California.

Eighty percent of the world's almond production comes from the Golden State. The climate is ideally suited to intensive production, so farmers exploit the area to the max. The long rows of almond trees cover an area of roughly 450,000 acres.

The almonds are harvested in September using a mechanical shaking machine that shakes each tree to make the almonds tumble down. They are then left to lie on the ground for a few days to dry before being swept together and sucked up by an oversized vacuum cleaner that drives between the trees. And now we're getting to the root of the problem: ideally, there should be nothing between the almond trees but bare, hard-packed earth, as this improves both efficiency and almond hygiene.

However, it also means that natural flower pollinators such as bees and other insects can find nothing to eat for miles around, which is not great, considering that almond trees rely on pollination to produce almonds. This explains the massive moving operation that takes place each February. The honeybees are needed on site, so more than a million beehives are transported on specially constructed trucks from all over the United States. The entire business is like a vast military exercise. More than half of all the beehives in the United States can be found in California every spring, just so that you and I can enjoy our marzipan.

So the next time you're chomping on a chunk of marzipan, send a few friendly thoughts in the direction of the bees—marzipan's midwives.

Beans, Bees, and Bowel Movements

Coffee has many functions. It can give you a boost during your break; the coffee machine is an indispensable hub of workplace socializing; and a cup of coffee is a crucial morning pick-me-up for many people, myself included.

Legend has it that an Ethiopian goatherd was the first person to discover the stimulating effect of coffee. He noticed that after eating the red beans of coffee plants his usually grumpy goats began to gambol around joyfully—and so did he when he tried them himself. One day, a passing monk managed to work out the connection. Hey, presto! He was suddenly able to stay awake through even the lengthiest prayer services.

Although this may not necessarily be the whole and unvarnished truth about the origin of coffee drinking, what *is* undeniable is that we are continually learning more about the role different animal species can play in ensuring that our beloved coffee finds its way to our cup. But we're talking here about animals that are considerably smaller—or a great deal larger—than a goat.

Let's start with the insects. It seems that even though the most common coffee plant can sort out pollination for itself inside each flower, the coffee crop is much greater if the bushes exchange pollen. And since the coffee flower blossoms for an extremely short time, nothing works better than a grain of pollen delivered straight to the door by express mail. Or straight to the stigma, to use the correct botanical term for the female part of the flower.

Who are the couriers? Many different varieties of bees. Studies show that bees can increase coffee yield by up to 50 percent.

In areas where the honeybee has not been introduced, more than thirty different species of solitary bees work amid the coffee flowers. Solitary bees are species in which every female takes sole responsibility for her own children—unlike social bees, of which most individuals are sterile but help bring up the queen's offspring.

Social bees such as honeybees are also good at pollinating coffee flowers, so coffee producers used to be encouraged to keep colonies of these bees near their coffee plantations. However, the prevailing view nowadays is that introducing honeybees may displace diverse solitary bees, who do a better job overall.

In order for solitary bees to thrive, it is important for there to be enough nesting places for them near the coffee plantation. Some species need small patches of bare earth for their nests, while others live in hollows of old or dead trees. The traditional method of cultivating coffee—in small fields of coffee bushes surrounded by woodland—is a far better way of ensuring pollination than doing so in a plantation in full sunlight. What's more, shade-grown coffee tastes better.

While we're on the subject of flavor, did you know that the world's ultimate luxury coffee is really crappy, in the literal sense of the word? When a coffee bean passes through an animal's intestines, some of the components are broken down and the coffee bean that emerges is sweeter and less bitter.

This remarkable discovery started with the Asian palm civet, a member of the Viverridae family. It lives in the tropical rain forests of Indonesia, where it enjoys a varied diet of small animals and fruit, including familiar exotics such as mango, rambutan, and—yes, that's right—the beans of the coffee plant. Don't ask me who, but somebody struck on the idea of extracting semi-digested coffee beans from the dung of the Asian palm civet and selling them for a stiff price. We're talking about more than $60 for a cup of civet coffee!

This started off as a nice sideline for small-scale Indonesian farmers, who began to collect wild civet muck. But once people realized there was good money to be made, they started capturing and caging the civets, which were then force-fed coffee beans under miserable conditions—an absolutely wretched business, to be avoided at all costs.

If you must insist on drinking expensive coffee, why not go for the elephant-dung variant instead? It's produced by a charitable foundation dedicated to the protection of elephants. Three days after a coffee bean enters at the trunk end, it is picked out of the dung at the tail end; the coffee made from it apparently has a sort of raisiny flavor.

Personally, I'd rather stick to straightforward shade-grown coffee with some raisins on the side.

Redder Strawberries and Tastier Tomatoes—Thanks to Insects

We have gradually come to realize that insect pollination is crucial for increasing the yield of crops of different fruits and berries. But did you know that insect pollination also helps improve the quality of the berries?

Take the strawberry, for example, which is not a berry in strictly botanical terms but a so-called false fruit—a swollen, juicy flower base covered in fruits (or actually nuts, botanically speaking, just to complicate matters further). The point is that each of the small pale "seeds" on the outside of a strawberry is actually a tiny fruit, and as many of these as possible must develop in order for the strawberry to become big and juicy. If only a few of the "seeds" develop, the strawberry will be small and knobbly. A well-pollinated strawberry may have 400 to 500 "seeds," and it takes insects to make that happen.

A German study shows that insect-pollinated strawberries are redder, firmer, and less ill shaped than strawberries that are wind- or self-pollinated. Firmer berries may not only be tastier but also cope better with transport and storage, meaning that they last longer in the stores, and that in turn means the strawberry farmer is better paid for the berries she has grown. The market value of insect-pollinated strawberries is 39 percent higher than that of wind-pollinated strawberries and as much as 54 percent higher than that of self-pollinated strawberries.

Similar effects are seen in numerous other insect-pollinated food plants. Apples become sweeter, blueberries grow bigger, rapeseeds have higher oil content, and melons and cucumbers are firmer fleshed. Even when gardeners have run around a tomato greenhouse with a vibrating wand that imitates the quiver of a bumblebee shaking loose the pollen, the taste panel is unanimous in its judgment: insect-pollinated tomatoes taste better.

Food for Our Food

But producing honey and ensuring pollination are not the only useful services insects perform for us and our food. They also form a vital part of the staple diet of the animals many humans like to eat, including larger species such as fish and birds.

Freshwater fish live largely off insects because some insects take infant swimming so seriously that they keep their young permanently submerged until they reach the age of reason: mosquitos, mayflies, and dragonflies, to name but a few. Many of the insect babies end up as snacks for trout and perch. And then we humans eat the fish. So you can thank insects next time you sit down to a trout supper.

Birds are also eager insect eaters. More than 60 percent of the world's bird species are insect eaters. Insects are particularly important as food for the chicks of some species, which are in serious need of protein-rich food to grow big and strong. Some of the birds we humans like hunting and eating, such as woodland birds and ptarmigans, are dependent on juicy insect larvae as chicks.

We humans can use insects as a source of nourishment, too. The United Nations estimates that more than a quarter of the world's population consumes insects as part of its diet. Insects are most commonly eaten in Asian, African, and South American countries, but Europeans also have a certain tradition of this. The Bible provides a precise description of which insects may be eaten—although the understanding of species isn't quite up to modern-day standards (insects have six legs, not four):

"All flying insects that creep on all fours shall be an abomination to you. Yet these you may eat of every flying insect that creeps on all fours: those which have jointed legs above their feet with which to leap on the earth." (Leviticus 11:20–21, New King James version.)

This is often interpreted as meaning that orthopterans are fine while other insects are unclean. And we know that locusts were considered a delicacy in ancient times: stone reliefs from around 700 BCE show skewered locusts being brought out for the king.

Insects Are Healthy, Environmentally Friendly Food

Insects are actually a very nutritious food. Of course, it depends on the type of insect, but generally speaking, they have the protein content of beef but contain very little fat. They also have many other important nutrients: cricket flour may contain more calcium than milk and twice as much iron as spinach.

And it isn't just healthy to eat insects, it may be environmentally smart, too. Replacing some of our current livestock with minilivestock such as grasshoppers or mealworm beetles could help make food production considerably more sustainable. For some people, it could also ease the transition to a diet based less on meat and more on plants.

Because, as we know, we're pressed for space on this planet. There are already more than 7 billion of us humans here, and 140 newcomers are added every minute. This is equivalent to a net monthly increase the size of the population of Maryland. And when it comes to putting food into all these mouths, insects are a much more efficient option than our traditional domestically farmed animals. It is estimated that, at their best, grasshoppers are twelve times as efficient as cattle at converting fodder to protein.

What's more, they consume only a fraction of the water a cow does and produce almost no dung—unlike cows, which really crap on the environment, so to speak. Cows produce tons of dung every year and emit a hefty dose of methane and other climate gases on top of that. There's very little of that kind of pollution in insect dung.

To cut a long story short: insect minilivestock requires very little space, food, and water, reproduces at a tremendous rate, and simultaneously provides an effective source of nutrient-rich food that is high in protein and emits minimal amounts of climate-changing gas.

Could it get any better? Well, yes, actually, it could. Because insects can also be raised on our food waste (see page 104). This means we can kill two birds (or grasshoppers) with one stone, producing good food and eliminating our waste problem at the same time. More research is needed into this area if we seriously intend to include insects in our diet or the diet of our livestock, and this is certainly a growing area of interest. One of the promising initiatives is to farm black soldier fly larvae on food waste. These maggotlike creatures are capable of transforming four times their own body weight per day into both feed and compost. Harvested in their last larval stage or as pupae, when their nutritional value peaks, they can be used as feed for fish, poultry, pigs, and even dogs. They are also edible to humans and can have a protein content as high as 40 percent. In addition, the leftover frass (leftover feed, exoskeletons, and droppings) from production can be used as fertilizer for plants. As almost one-third of all food produced for human consumption is lost or wasted every year, according to the Food and Agriculture Organization (FAO), the potential for alternative solutions here is certainly enormous.

Bug Grub

If using insects for human consumption is to yield any environmental benefits, it won't be a matter of scattering a few fried ants on our salad or decorating confectionery with chocolate-coated grasshoppers. Those stunts where cooks serve up whole insects are mostly just for novelty's sake.

Because just as we don't eat lamb chops with the wool on, insects will need to be processed to transform them into tempting foodstuffs. And we must produce them in quantities that make the finished product cheap and easily available. Only then can cricket flour protein cakes and mealworm burgers become daily fare.

The term "Meat-Free Monday" has already caught on in some places. Maybe "Insect Tuesday" will be next.

It may be some time before we start thinking of insects as everyday food. But how about developing insect-based fodder for livestock or fish in the meantime, using insects that have eaten their way through our organic waste en route to adulthood? That way we could feed our farmed salmon insects instead of soy. Research on such topics is happily already under way.

Using insects as food for humans involves some challenges. Insects have their share of parasites and diseases, which we will have to control if we are to engage in large-scale production. Some people have allergic reactions to insects, and legislation dealing with insects for consumption will need upgrading.

Crucially, we must be certain that this is genuinely sustainable, from a life-cycle perspective, too—that heating the mini–cattle sheds doesn't eat up all the benefits, for example. Because grasshoppers aren't like hardy sheep breeds that can be kept outside all year: they won't necessarily be able withstand to the climate throughout the year without heating, especially in northern climes. Proper heat levels are crucial to rapid growth and high reproduction.

One major challenge remains: consumer acceptance. Consumers must want to buy and eat insect products because they see them as interesting and useful food. Perhaps this will come about naturally if cheap and tasty insect flour becomes easily available in the store. We can do this if we want to. It took us only a few years to learn to eat raw fish, after all. Could insects be the new sushi?

It is also important to find the right way to describe these new tidbits. Grasshoppers and crickets, which people already eat in many other parts of the world, could, with a little imagination, be repackaged as the land's answer to shrimp. And we have to use language that sparks positive associations.

Pioneering promoters of the insect diet are on the case. And in the English-speaking world at least, humor seems to be the answer.

Grub Kitchen, a restaurant in Wales, uses punning language to help make the food more palatable—from its name to the often alliterative offerings on the menu: mealworm macchiato, bug burgers, cricket cookies. Its chef, Andy Holcroft, has also pointed out that the language used about the dishes must appeal to the senses: "Crispy and crunchy descriptions of insects, such as stir-fried or sautéed, sound more appetizing than soft-boiled or poached . . . [which sound] squelchy and squishy." In Norway, the word *mushi* has been suggested to promote the softer insect dishes, both to draw on the positive associations of *sushi* and because it means "insect" in Japanese.

If You Can't Beat 'Em, Eat 'Em

The nineteenth-century British entomologist Vincent M. Holt was extremely interested in nutrition, especially among Britons of limited means. He thought the working classes should turn to insects as a rich source of nourishment. As early as 1885, the same year in which the Statue of Liberty arrived in New York and Henrik Ibsen's *Wild Geese* had its world premiere in Norway, Holt published a neat little book entitled *Why Not Eat Insects?* For the occasion, slugs and snails (which are mollusks) and wood lice (which are crustaceans) were incorporated into the insect world.

Holt argued forcefully that insects would be a healthy and useful addition to the menu. He thought they would spice up the wretched diet that the working classes and farm laborers ate in those days. The peasant should eat the pests in his field for dinner, and the woodman should lunch on the fat larvae in the trees he chopped down, Holt suggested—a win-win situation, in other words.

Holt's funny little book also includes plenty of recipes. Unfortunately, perhaps, neither his Slug Soup nor his Fried Soles, with Wood-louse Sauce quite caught on. Perhaps a better choice of raw materials and modern methods of preparation might help combat the lack of enthusiasm for eating insects. Nowadays the matter is under serious discussion at the United Nations as well as other forums.

Perhaps the future will prove Holt right in the end: "While I am confident that they will never condescend to eat *us*, I am equally confident that, on finding out how good they are, we shall some day right gladly cook and eat *them*."

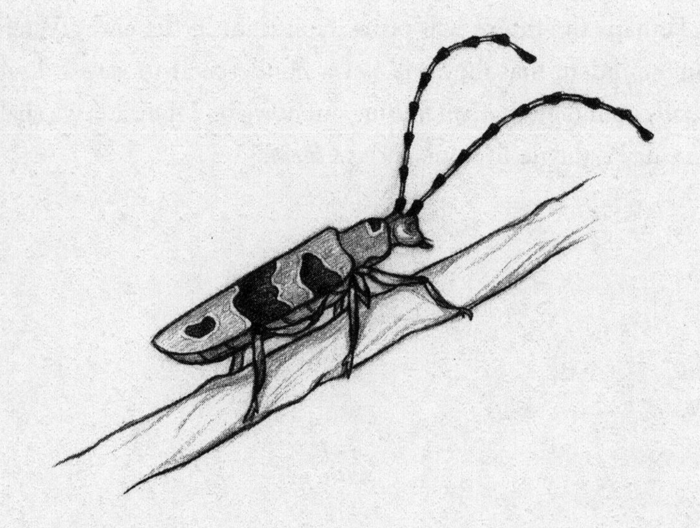

6

The Circle of Life—and Death

Insects as Janitors

I know of few things as beautiful as mighty, ancient oaks. There they stand proudly, a legacy from a bygone age; oaks that sprouted and grew before the days of streetlamps and social media, in a time when trolls still lived among the trees and not in the shimmering websites on your computer screen.

Today's great oaks have retained their magic. And where Pippi Longstocking once found lemonade, we scientists can go in search of rare insects. Because inside the ancient oak trees, hollows form where the wood slowly rots. Inside, it is dim but not quite dark. There is a scent of fungus and damp earth, like a faint suggestion of autumn. At the same time, a sweetish hint of warm timber is like a promise of spring to come. Here you discover another world, in which the meaning of time and space is altered. Time goes faster because a beetle lives its entire life over a single summer. And a fistful of reddish brown wood mold, with its raw tang of fungus, damp, and life's decay, is an entire world for a millimeter-long pseudoscorpion.

Inside live brightly colored red velvet mites and pallid beetle babies, enormous scarabs and tiny springtails. Nurseries and pickup joints stand side by side. There are life and death, drama and dreams, all on a millimetric scale.

The quest to find ancient oaks and their inhabitants has led me to many forest areas I would never otherwise have seen and given me many encounters with nature that I wouldn't have missed for the world: picnic spots on bare, rocky hills with blue mountains in the distance and the warm spring sun upon my face; late spring evenings in Telemark on my way back to the car after my day's work is done with only the cry of the tawny owl and a sickle moon for company; steep slippery slopes that I barely managed to scale in pouring rain; boulder scree in western Norway, where all the oaks bear traces of pruning from earlier times, when people harvested leaves as winter fodder for their livestock; avenues, pastures, wooded hillocks in cultivated fields, private gardens. Usually solitary—yet never alone. Because more individuals may live in these ancient oak trees than there are human inhabitants in Oslo.

An ancient hollow oak tree is like a fortress—a fortress of biodiversity, no less. The shell of resilient oak wood provides shelter from rain, sun, and hungry birds for the many hundreds of different insect species that live inside. The intricate oak bark, reminiscent of the ornamental carvings of intertwining dragons and serpents on Norway's stave churches, provides a habitat for minuscule pin lichens. Some fungi live in close cohabitation with the roots of the oak, while others help the insects break down the dead timber.

The factor largely responsible for the presence of all these species is wood mold, a life-giving blend of rotting wood residue, fungal threads, maybe an old bird's nest, and a smidgen of bat guano. For insects, wood mold is like a gourmet restaurant: even the most exacting bugs will find a menu to suit their tastes. Hundreds of different insects may live in the dim, humid atmosphere inside a hollow oak, contributing to nature's eternal circle by slowly converting mighty trees into mold and soil where new acorns can sprout.

Somebody Has to Do the Cleaning

Herbivores eat just a tenth of all the plants that sprout and grow. All the rest, 90 percent of all plant production, is left lying on the ground. Plants and trees aren't the only things to die; life also comes to an end for animals of all sizes, from midges to moose. As a result, there are impressive amounts of protein and carbohydrate to be recycled. On top of that we have to consider all the waste these animals produce over their lifetimes; dung, plain and simple, needs to be taken care of, too—a pretty crappy job, you might think, but insects are on hand to help us out, as usual.

This is where nature's caretaking service comes in. Just as in schools, offices, or apartment buildings, it's often the janitor who must clear up after everybody else. And this is how it works in the forests, the meadows, and our cities, too, where thousands of fungi and insects perform the crucial task of decomposing dead organic matter. Nature's tiny janitors devour the mess on the spot. It can take time, and it requires a sophisticated collaboration in which different species have different roles to play.

And even though very few of us think about it as we take a

Sunday stroll through the park or in the forest, these processes of decomposition are crucial to our life on Earth. Insects' patient chomping on dried-up trees and rotten remains doesn't just clear the ground of dung and dead plants and animals; just as important, the insects' contribution returns the nutrients in the dead organic matter to the soil. If substances such as nitrogen and carbon are not returned to the earth, it will be impossible for new life to grow.

Dead Trees as Beetle Abodes

When the insect mom is house hunting in the forest, her priorities are different from ours. Take beetles that live in dead trees, for example: whereas we fear moisture damage and rot, beetles think they're fantastic because they're like a fridge full of food for the family's greedy kids.

So Madame Beetle goes for a viewing. Softly, she sets down all six legs on the dead tree. With antennae and toes, she tastes and smells the spot where she's landed to see if it will make a good nursery for her beetle babies. If she's satisfied, she swiftly lays her eggs in a little crack in the bark and moves on in search of more trees in need of caretaking services.

Each egg hatches a larva that boldly sets about chewing its way through bark and wood—a gargantuan task in which, fortunately, it is not alone. Thousands of beetle larvae may be at work in such dead trees, ably assisted by bacteria and fungi.

New dead wood is fantastic fun: there's plenty of sugary sap beneath the bark, and when it ferments there's a real party atmosphere among the visitors. Every type of wood has its beetle specialists, who greedily gorge on this delicacy. Bark beetles are a typical example. But speed is of the essence, because by the end of the first summer the platter is empty: all that lovely sugar is gone.

By contrast, dead, dry timber is a pretty dismal lunch for a beetle. Cellulose and lignin, two of the most important components of timber, are about as juicy and digestible for an insect as a sack of bran is for us. So it's a good thing that some fungi just adore cellulose and others love lignin. They cast their fungal threads—bits of fungus they carry on their bodies or in their guts—into the wood, making it more appealing to the beetles because the nutritional content increases and becomes more accessible. Bacteria also provide a great garnish. Some beetles even have tiny collaborating partners in their bodies that help them extract nourishment from even the most indigestible parts of the tree. All in all, myriads of organisms are involved in the decomposition of dead wood.

Dead Wood Lives!

Dead trees, branches, and roots are home to a surprisingly large number of species. As many as 6,000 species live in dead wood in the Nordic countries—a third of all the species found in our forests! Approximately 3,000 of these are insects. By comparison, there are only around 300 bird species and fewer than 100 mammals in the region.

Once fungi and insects, mosses and lichens, and bacteria have moved in, there are more living cells in the dead tree than there were when it was alive. So ironically enough, dead trees are actually among the most living things you can find in the forest. And every species has its particular cleaning job to do, not to mention having its own precise demands when it comes to the kind of dead wood it wants to live in or on.

Why are there so many species in dead wood? Part of the reason is that insects that live on dead timber have differing demands when it comes to the type of dead wood they're after. For us humans, who don't find timber toothsome, it is difficult to grasp all the nuances when it comes to types of wood, stages of decomposition, size, and surroundings.

But to insects, a dead spruce tree is quite different from a dead birch. And an aspen that has just died is quite unlike an aspen that has lain dead in the forest for several years. As I mentioned earlier (see page 94) plants and trees have an active, species-specific defense against grazing animals and insects. This persists after the tree's death, especially early on, meaning that the first insect arrivals at the newly dead tree must be specially adapted to cope with this.

Size is also important: a dead oak branch offers a totally different habitat than the rotted innards of a giant oak. And a dead resinous pine tree on a hill in the baking sunlight is home to totally different species with different dietary habits than a dead pine in a dark, dense forest.

In other words, a stick isn't just a stick; dead timber has more nuances than a fine wine, and many insects are exacting connoisseurs. Since insects have such differing demands, there needs to be enough of all these various types of dead wood in the forest to provide enough housing to enable every insect to find its own little hovel and get its job done.

But there's one more vital point when the beetle mom is hunting for a suitable dead tree trunk in which to house her children.

The window of opportunity is short in the insect housing market, and it's a question of getting to the suitable tree in time. If the distance between rare and specific types of logs (such as large-diameter oak trunks) is too large, as is often the case in modern managed forests, the beetles dependent on such logs might not be able to get there at all.

That is why natural forest, forest that isn't affected by modern logging, is so important. It contains much more dead timber than managed forest, and there is a great deal more variety in the dead wood, which means there are more beetle homes on the market. They are so close together that the beetle mom can pop into several of them in a single evening of viewings, laying a few eggs here and there. That's the way to create beetle diversity.

Bang-up Research

The goings-on in dead trees are one of my favorite topics and a subject we do a lot of research into in the research group I belong to. It may not all be "rocket science," but we've certainly been involved in some projects that went off with a bang. Such as fifteen years ago, when we did a really bang-up experiment: we wound yards of detonating cord around trees in the forest, 15 feet above the ground, and lit the fuse. And then we ran . . . With an enormous bang the trunk was blown right off, and the top of the tree crashed to earth!

The point of the exercise was to create standing dead trees. We created sixty such trees, and in each of the years that followed we checked which beetles were visiting them. This taught us a great deal about different insects' dietary preferences. We also saw that the forestry sector's environmental measure of tree retention—leaving trees in the clear-cuts that eventually become high stumps—actually works.

What's even more fun is that now, fifteen years later, we can hear a kind of echo from those earlier beetle visits. It turns out that there are different fungi on the trees these days depending on which insects paid a visit all those years ago!

That made us wonder whether fungi and beetles have become a bit like bees and flowers—are they mutually useful? Perhaps certain wood fungi simply hitch a ride with certain beetles and get dropped off at the restaurant door? We know of some bark beetles in which the collaboration is so close that both parties depend upon it. But could this collaboration be much more common than that in a looser form—without codependency but with advantages for both parties?

To check this out, one of our PhD students has been putting trees into cages—or rather, parts of trees. She cut down living trees, made logs of identical sizes from them, and randomly chose which logs to cage. Logs that were caged did not receive insect visits, as the bugs couldn't get through the netting of the cage. For the purposes of comparison, the other logs were placed outside the cages, so that insects could land as usual.

It turned out that the fungal society was totally different in the logs that insects couldn't access. We think this is because many insects carry spores or fungal threads, in their body or their guts. When the insect lands on a newly dead log to lay eggs, this fungus is sprinkled or excreted along with the insect dung and thereby finds a new home.

In addition—and this is the really exciting bit—our study showed that the caged logs decomposed more slowly. The cleaning job takes longer when the insects don't get to help.

A Kindergarten beneath Your Shoes

I love running, especially on soft forest paths. A half-hour run from my home takes me to a forest reserve strewn with dead trees, like a game of pickup sticks. I can look around and try to count species, of which there are around 20,000 in Norwegian forests. Of course, not all of them live in "my" forest, but still—how many can I see? I can count several trees, a dozen plants, lichen, fungi, perhaps an elk or a large bird if I move quietly. If it's summer, the insects will do wonders for my list of species, but even so, I spot barely more than a hundred. Even here in the reserve. So where are all the other tens of thousands of species?

A lot of the other species are tiny insects and related creatures that live out their lives in hiding. As I mentioned, a third of forest species live in and on dead trees. The other important habitat is the soil; there is no other place where species are packed together so densely. The tiny patch of earth stuck to the sole of my running shoes after a trip to the forest may be home to more bacteria than there are human beings in the United States, not to mention thousands of thin fungal threads. Here in the soil are also found a myriad of important critters and small insects. A whole zoo of little creatures lives down there in the darkness: earthworms and mites, roundworms, pot worms, springtails, and wood lice. All these species, which we don't give a hoot about on a day-to-day basis, have important jobs in the recycling sector. They chew and dig and aerate and mix. In the blink of an eye, rubbish is reconverted into soil, ready to sprout new life. It's pretty miraculous, really.

Soil is important, but masses of it vanish every year—not because runners dash off with chunks of it stuck to their shoes but because of erosion by wind and water. Some of this is natural, but in many places the soil loss is high because we humans have removed natural vegetation. Consequently, there's nothing left to retain the soil, which is blown away or runs off into the sea and elsewhere. This is losing us billions of tons of topsoil every year—and along with the soil, we are losing the diversity of decomposers, which are our guarantee that the recycling of nutrients will continue.

The thin layer of earth is the planet's skin, a thin living layer over the magma and the rocky crust. Perhaps we ought to pay a little more attention to the earth's skin care? Like a teenage girl anxiously checking her complexion in the mirror, we, too, should be aware of the well-being of the topsoil and forest soil, along with all their inhabitants. Because we need them and—to stick to the language of the cosmetics industry—because they're worth it.

An Ant in Manhattan

Finger food at festivals, picnics in the park. The summer propels us and our meals out into the city. But what about the scraps of food—the bits of hamburger we drop on the pavement or the hot dog bun left lying on the grass? This is where ants come into the picture.

Many people think of ants as a nuisance, even yucky. But it's actually good to have them around, in the urban environment, too. A group of insect scientists studying ants in Manhattan did a back-of-the-envelope calculation and estimated that there are two thousand ants for every person in the city. And what do ants do there? They live out their tiny ant lives, which consist mostly of food gathering and reproduction. They are undemanding when it comes to diet and have a healthy appetite. Another of the scientists' back-of-the-envelope calculations estimates that the scraps of junk food put away by the ants of Manhattan add up to the equivalent of sixty thousand hot dogs a year! It's a damn good thing we have them.

In an experiment, scientists compared how much food waste found its way into ant bellies in different parts of Manhattan. Precisely weighed quantities of food were set out in tiny "food scrap cafés" in parks and on pedestrian traffic islands. The scientists offered the ants a comprehensive fast-food package, New York style: hot dogs, and a cookie for dessert. At the same time, they measured the species richness of ants and other urban bugs in the same places, establishing that there were more species of ants (and other bugs) in the parks than on a busy street.

Because it has been shown that food gathering in many other natural systems is more efficient in a species-rich society, the scientists expected the park ants to eat more of the food scraps than the traffic-island ants did. But the result they got in Manhattan was the opposite: the ants on pedestrian islands carried off more than twice as many of the scraps. There may be several reasons for this. First off, it is warmer on the pedestrian islands. And since ants are cold-blooded animals, everything goes faster when the temperature is high.

Second, it seems that a European immigrant, the pavement ant, really has a taste for American junk food. This species was much more common on pedestrian islands than in the parks, and wherever it was present, up to three times as many fast-food scraps vanished as when it wasn't. In other words, environmental conditions and individual species turned out to be more important than species diversity when it came to cleaning up Manhattan's food scraps.

Pavement ants are territorial, and, like other urban gangs, they fiercely defend their little patch of the city against interlopers. But the ant gang isn't alone on the streets of Manhattan. There are regular episodes of gang violence between rats, which are less common but bigger. They want their share of the junk-food plunder. These clashes between gangs ought to be of some interest to us humans, who are even bigger. Because although mice and rats make a positive contribution in the sense that they eat our food scraps, they are also notorious transmitters of disease. The same can hardly be said of ants. This means that ants are much better suited to the role of cleanup patrol in the city's outdoor spaces.

It is time to acknowledge that even our cities are ecological systems, in which crawling critters are an essential ingredient. On a pedestrian island on Broadway alone there are thirteen different ant species. In all, forty ant species have been found in New York; that's almost two-thirds of the ant species in the whole United Kingdom. And since more than half of the world's human population now lives in cities, we ought to spend more time finding out how urban ecosystems operate.

The thing is, urban nature also performs significant services for the ecosystem. Trees provide shade, muffle noise, and clean the air. Green areas absorb water after heavy rainfall and reduce flooding. Open water cools the air, and the species in ponds and streams filter the water, making it cleaner. The tiniest patch of earth can provide a habitat for masses of useful bugs that pollinate plants, spread seeds, or clean up the streets—such as ants.

Economists in Oslo have studied that city's ecosystem services and their worth. An attempt to measure the value of the green structures in and around the capital to the health and well-being of the inhabitants, for example by calculating their use value, among other measures, added up to millions of dollars. And that was without including the value of the ants' contribution.

A greater knowledge of urban ecology will enable us to plan and maintain our cities better. Even something as simple as raking pedestrian islands less frequently has proven important, as it ensures more hiding places and a happier life—if you happen to be an adventurous Manhattan ant.

A Troublesome Fly

Hot dogs on the streets of big cities are one thing. But there are also other types of dead meat that need to be cleared away out in the natural world. Think of all the animals, big and small, that die and are left lying where they dropped. It would be quite unpleasant if they didn't get recycled pretty quickly.

From the insects' point of view, carcasses are a handy source of food—they can't run away, and they can't defend themselves. But the insects have to be quick, because carcasses are rich in nutrients and therefore much-sought-after food; what's more, the competition includes a large range of species of varying sizes. Here, insects are literally in the flyweight class, whereas their opponents are heavyweights such as foxes and ravens, vultures and hyenas. One trick is to lay not eggs but ready-hatched larvae in the carcass, as some flesh flies from the Sarcophagidae family do. Another is to eat quickly, grow even more quickly, and generally be flexible when it comes to how big you need to be before pupating.

Another cunning solution is to hide the carcass by burying it. The beautiful red-and-black burying beetles of the *Nicrophorus* genus are masters of such a vanishing act. They work in pairs, digging out soil from underneath the carcass and placing earth on top, and in this way they can bury a dead mouse in a single day. Beneath the earth, they wrap up the carcass in a ball and lay their eggs on it. And despite the slightly jaw-dropping choice of nursery, they are attentive parents: they chew off tiny scraps of the carcass and regurgitate them into the mouths of their larvae, which are incapable of digesting the food themselves. This is one of the few examples of parental care in the insect world other than among the social insects (see page 44).

Burying beetles also have some good friends that are not insects. When newly hatched sexton beetles leave their childhood home, masses of tiny mites climb onto them and hitch a lift to the next carcass. This species of mite lives only with burying beetles; it cannot fly and is reliant on transport to find its way to a new, fresh carcass. In return for the lift, the mites eat up the eggs and larvae of other competing fly species in the carcass.

The decomposing crew that turns up to break down the carcass belongs to a segment of the insect world that is rarely mentioned or rewarded. There are no fan groups for burying beetles as there are for bumblebees. Yet they are tremendously important animals.

In South Asia, people have learned to their cost what can happen when carrion eaters vanish. It's true that the animal in question was the vulture, which might be said to be the blowfly's massively big brother and enjoys a similarly bad reputation among most people. Nonetheless, the point is the same. Around the turn of the millennium, the veterinary medication diclofenac was introduced into India as a treatment for sick cows. Just fifteen years later, the medicine had dispatched an insane 99 percent of all the country's vultures because residual amounts of the substance were left in dead cows and were passed on to the vultures that ate them. The vultures suffered kidney failure and died. Although scavenging insects almost certainly worked at top gear, they were unable to deal with such large amounts of carrion alone. As a result, dead cows were left lying on the ground. Once the vultures had vanished, other large scavengers appeared on the scene: feral dogs, whose numbers exploded. Since many of them are rabies carriers, the population boom in dogs that resulted from the disappearance of the natural carrion eater has been blamed for an additional 48,000 rabies deaths in the Indian population.

In fact, carrion eaters can also help the police in criminal investigations. There is a pattern to when which species come to a corpse, and this can be used to help connect the dots in criminal investigations and ultimately solve crimes. The first time insects helped identify a murderer is supposed to have been in a Chinese village in 1235. A man was brutally murdered with a sickle, and the local peasants were called into a meeting. They were instructed to bring their sickles with them. The investigator made them wait, and, since it was a hot, sunny day, it wasn't long before flies appeared. When all the flies landed on the same sickle, the owner was so shocked that he confessed on the spot. With their peerless sense of smell, the flies were drawn to the traces of blood even though the sickle had been cleaned.

Today the methods are more advanced, but the basic principles remain the same. Insect species appear in a dead body in a given order and following a particular logic. This fact can be used to calculate the time of death and may, in some cases, also tell you something about the cause of death. Drugs and toxins accumulate in the insects present, and thus they can be more easily detected. Such chemical substances also affect the growth rate of feeding maggots and is therefore important information when forensic entomologists are estimating the time passed since death occurred.

Moreover, species are distributed over defined geographies. Knowledge of this fact can be used to determine whether a body has been moved if the species that are present are normally widespread in other areas or in other parts of the country. One example of this was a case where a body was found in a sugarcane field in Hawaii. The oldest larvae that were found in the corpse belonged to a fly that lives primarily in urban areas. And indeed it turned out that the corpse had been kept in an apartment in Honolulu for a couple of days before being dumped in the field.

Insects can also make a more indirect contribution to crime solving. Insects mashed into a car's radiator grille were used to trap a killer in the United States. He claimed to have been on the East Coast when his family was murdered in California, but the species found on his rental car could be found only on the West Coast.

When Nature Calls and Insects Answer

All animals eliminate waste in the form of dung. Dung from large animals, like mammals, represents a significant biomass. Dung may contain useful nutrients, but it also contains large quantities of bacteria, disease-causing parasites, and other things the body has expelled. Not all animals are up to eating this waste, but insects stand ready. Beetles and flies are particularly likely to have muck on their menu. In this working group a special expertise is required: a good sense of smell and rapid reflexes. When the battle for cowpats is under way, you have to be quick if you want to secure your slice of the pie.

Some participants, such as the horn fly, are renowned for starting to lay their eggs before the cowpat is even finished. This is dangerous, but some parents will go to any lengths to ensure the best conditions for their kids to grow up in—because fresh dung goes quickly, especially if it's warm; you might say it sells like hot cakes. For example, one study showed that as many as four thousand dung beetles fell upon a 17-fluid-ounce dollop of elephant dung doled out by researchers in a mere fifteen minutes. Other studies found that it took a couple of hours for 3 pounds of elephant dung to vanish once 16,000 dung beetles went in and did their job.

Dung beetles have three main strategies: they can be dwellers, tunnelers, or rollers.

The dweller likes to live right in the middle of the meal. It crawls down into the dung, content to eat and lay its eggs there. Many Norwegian dung beetles (members of the Aphodiinae subfamily) belong to this category. The dwellers' strategy is risky. They never know how many others are laying their eggs in the same dung, and in the worst case the larvae may eat one another out of house and home and all end up starving to death.

One way of avoiding this is to build an extension for the kids, with its own pantry. This is the technique used by the tunnelers. They dig passages beneath or right beside the dung that can range in length from four inches to a yard long. We often find the longest passageways in species where the dung beetle mother and father work together. They drag small balls or rolls of muck down into their tunnels, which then serve as a kids' room for the larvae.

The most advanced variant is the rollers, which grab their share of food and make a swift exit. They pack the dung into a ball, which can often weigh fifty times as much as the beetle itself, and trundle it off—always in a straight line, regardless of whether the sun is hidden behind a cloud or whether it's a dark, starry night. How do they do it?

Creative scientists have really gone to town on their field experiments: Some placed tiny peaked caps on the beetles' heads to shade them from the sun. Others used large mirrors to manipulate the position of the sun or the moon. The most creative of all was, perhaps, the researcher who moved the entire experiment into the Johannesburg Planetarium and proved that dung beetles can use the Milky Way to orient themselves! The only other animals known to use the stars for orientation are human beings, seals, and some birds. All in all, the research shows that the dung rollers can steer their course using the positions of both sun and moon, as well as polarized light or the Milky Way.

These particular beetles have fascinated humans for thousands of years. The dung-rolling *Scarabaeus sacer* played a central role in Egyptian mythology. When the Egyptians saw those beetles trundling off with round balls of dung, it reminded them of the sun's journey across the heavens. The beetle became their "sacred scarab," symbolizing Khepri, the god of the rising sun. This insect god is sometimes portrayed as a beetle and sometimes as a man with a beetle's head.

The Egyptians also saw that the scarab beetles were among the first living things to emerge onto the muddy banks of the Nile after the spring floods. Where the old dung beetles had buried their balls of dung, new young beetles clambered up out of the earth a few weeks later. From there, it wasn't much of a leap to make a link between the sacred scarab and renewal and reincarnation. It became common for scarab amulets to be used by the living and to be bound into the bandages that swathed mummies.

Perhaps the Egyptians even got the idea of mummification from beetles, because what does a beetle pupa look like, if not a mummy? It has even been suggested—somewhat playfully, perhaps—that the pyramids are sacred representations of piles of dung, in which the dead pharaoh lies like a mummified pupa, waiting for the metamorphosis of reincarnation.

Dung Does So Much

Dung can be used for so many things. In a lot of cultures, dried cowpats, for example, are still used as fuel or building material. In the insect world, too, we see examples of creative use of excrement. How about a dung wig, for example? The *Hemisphaerota cyanea* leaf beetle lives in dwarf palms in Florida and the Southeast states. As a larva chews its way through a palm leaf, beautiful pale yellow curly threads ooze out at the opposite end. The larva arranges these pale threads of dung neatly over its back until it ends up with a whole wig—not unlike Donald Trump's hairdo.

The point of the wig is, of course, self-defense: no matter how hungry you are, you're unlikely to fancy a mouthful of hair.

Several leaf beetle larvae adopt similar techniques, but instead of using hair, they count on intimidating and scaring off the enemy. The pale green *Cassida viridis* is common in Europe. Its larvae fashion a kind of roof or parasol out of old larva skin and black lumps of dung that they hold over themselves with the aid of a special "anal fork." If an enemy comes too close, the larva can brandish its dung parasol, which may also contain poisonous substances that the larva has produced from the leaves it eats in order to keep enemies at bay.

Case-bearing leaf beetles (from the Cryptocephalinae family) are even more advanced. Their children are equipped with something akin to a mobile home made of poo: The mother lays each egg in a beautifully formed container that she molds from her own excrement. When the egg hatches, the larva opens the door and sticks out its head and legs so that it can carry its house wherever it goes. As the larva itself defecates, it adds to its mobile home, thereby ensuring that it is always big enough. When it is time to pupate, the larva climbs inside and shuts the door behind it. There it can lie, nice and safe, until it has become an adult beetle—and the whole sequence starts all over again.

An Ecosystem in a Pelt

Some people think sloths are cute. You know, sloths—the creatures who were portrayed as the incredibly slow but smiley clerks in Walt Disney's *Zootopia*. I actually got up close to a sloth one time in the wild. And I didn't find it in the slightest bit sweet.

I was sitting on the outskirts of a village in Nicaragua with my back to a fallow area—half-open forest land with bare earth. It was pouring rain. I heard a noise behind me and turned toward the forest. There, just a few yards away—slowly, slowly, its gaze fixed firmly on me—came the most peculiar creature I had ever seen, slinking toward me, sopping wet. This was thirty years ago, but I clearly remember thinking: Good God, it looks like a nuclear mutant!

One biology degree and many years later, I realized this must have been a rare sight. Sloths are one of very few genuinely tree-living mammals and spend the absolute minimum of time on the ground. But once a week the time comes for them to do their business, and, oddly enough, they have to do it on the ground. That's when they tend to die, because they are so incredibly slow and can barely defend themselves from predators.

The last thing that occurred to me was to count the toes on the forepaws of this slightly terrifying creature with its fixed grin. Now I know that there are two groups of sloths, three toed and two toed, and a few different species within each group. The two- and three-toed variants are very different. It's the three-toed kind we'll be dealing with here.

Neither did it occur to me to go over to the animal and search for moths in its brownish green pelt. I regret that now because sloths carry an entire ecosystem in their fur—a fact we have only recently grasped.

Why should three-toed sloths risk going to the toilet on the ground instead of just letting rip from the treetops instead? Especially considering that they expend 8 percent of their daily calorie intake on these climbing trips and risk getting eaten to boot. Scientists have long sought an explanation. Could the point be to provide manure for the tree they are living in, or are they communicating with other sloths via their latrines?

That isn't it. In the pelt of the three-toed sloth lives a creature entertainingly known as a sloth moth. When the sloth goes for a toilet break, the moth climbs out of its pelt and lays some eggs in the poo. The larva lives there happily, and when it grows into an adult moth, all it has to do is wait for the sloth's next loo stop to move into a safe, warm sloth pelt.

And this is when the fun really starts, because surely the sloth wouldn't bother to risk its life just to do a moth a favor? Well, it turns out that this business involves some advantages for the sloth, too.

The moths excrete, die, and decompose in the pelt. This increases the nutritional content of the pelt and improves conditions for a type of alga that grows on sloth hair (and nowhere else in the world, just so you know). The sloth eats this green algal growth by licking it off its fur. The alga has a crucial advantage: it contains important nutrition that the sloth cannot get from its monotonous plant diet. It can also serve as camouflage.

So to sum up: the moth is good for the alga, the alga is good for the sloth, the sloth is good for the moth. It's a whole tiny ecosystem—all in one animal pelt.

Other large animals also play host to dung insects that have found it wisest to stay close to the source rather than spend their entire lives looking for fresh muck. Among kangaroos and our hairy ape brothers, certain beetles set up house in the pelt close to the animal's rear end. So there are a few benefits to having a hairless rear that probably hadn't occurred to you.

Drowning in Dung

In 1788, the first cow to set its four feet on Australian soil arrived. Along with it came a rather motley crew of 1,480 men, women, and children—mostly convicts—along with 87 chickens, 35 ducks, 29 sheep, 18 pheasants, and various other things. That marked the end of the Aborigines' 40,000 years of isolation, not to mention that of the animal and plant life, which had been isolated since the Australian continent split away from Antarctica somewhere between 40 million and 85 million years before. Consequently, the continent was full of species that existed nowhere else on the planet; 84 percent of the mammals and 86 percent of the plants in Australia were unique.

The four cows and two oxen that accompanied the first European fleet had been picked up during the crossing. They came from Cape Town and were zebus, a race of cattle that is accustomed to a hot climate. A convict named Edward Corbett was given the job of herding the animals, with strict instructions not to let the cattle out of his sight. But alas, just a few months after Daisy had ambled down the gangplank, she and the other cattle vanished. They'd taken off while the herder was having supper.

It was a minor catastrophe. The six cattle were supposed to be used for breeding, milk, and food. The settlers could find no familiar edible plants in Australia. Even though they had grain for planting, many of the prisoners had no experience with agriculture and weren't especially keen to learn. They weren't even any good at fishing. The provisions disappeared fast despite ultra-strict rationing.

So there was great rejoicing when they found the cattle again a few years later—by which time they had become a whole herd. They were getting by very well in Australia's pastureland.

After one or two hundred years, joy turned to desperation. Because what do cows do? They feed, chew, belch, and defecate, and they do all of these things on a massive scale. One cow produces as much as nine metric tons of dung per year, and that's dry weight. The excrement of a single cow covers an area the size of five tennis courts each year. And when cows thrive, there get to be a lot of them—with equivalently large numbers of tennis courts' worth of dung.

By around 1900, there were more than a million head of cattle in Australia. But who was going to clear up the crap? This brings me to the point of the story: there were no beetles in Australia that could decompose the cow dung. There were some native dung beetles, but they had been reared on dry, hard marsupial dung for millions of years. They had very little taste for foreign cuisine in the form of the zebus' mushy manure.

Therefore, the dung was left lying on the ground. There it dried into a crust that could not be penetrated by so much as a blade of grass. At the height of the problem, up to 500,000 acres of grazing land per year were becoming unusable. Around 1960, roughly two hundred years after the arrival of the first cattle, large areas of the country lay fallow because of dung that had not decomposed.

The only things that could be bothered with the dung were the flies, but they didn't exactly help. In Australia there's a type of fly that resembles the European housefly, except that it lives anywhere *other* than in houses—and now it was particularly keen on living wherever dung happened to be lying around unused. This troublesome fly bred massively, as did other flies that plague humans and livestock—an increased nuisance factor on top of the problem of large areas of fallow land no longer suitable for grazing.

New beetles had to be brought into play. A large-scale project was set into motion, sponsored by the government and the meat industry. Over a period of fifteen years, Australian entomologists experimented with numerous species, and after careful testing they set out 1.7 million individuals from forty-three different dung beetle species.

The project was a success. More than half of the species became established. The dung disappeared, and the plague of flies dwindled notably. Before, only a tiny share (15 percent) of the nitrogen in the cowpats was being returned to the soil; the beetles' caretaking service increased the level to as much as 75 percent. This example shows just how important decomposition is for nature and for us humans.

Despite their importance, things aren't looking so good for dung beetles as a group. Globally, 15 percent of the species are threatened. In Norway, more than half of the roughly seventy types of beetles that live in dung are listed as threatened with or near extinction, and thirteen species have apparently already vanished from the country. Norwegian dung beetles are having a particularly tough time in the south of the country—which provides a habitat for species that require fresh cow muck, preferably on sand or unfertilized pastureland beneath a warm summer sun. Changes in agriculture are largely to blame for the disappearance of the dung beetles. Uncultivated pastureland becomes overgrown or is not grazed continuously over time.

Another problem is the widely used antiparasite treatment ivermectin, which is given to cattle and other livestock all over the world. The substance has been found to be excreted in the dung in large amounts and to harm the dung beetles that come to clean up. This may have consequences for both species diversity and speed of decomposition. To reduce the negative effects on dung beetles, it has been suggested that the drug be distributed only by injections to reduce the amount excreted in dung, and only to animals where parasitic infections are severe. Such restrictions could also help in delaying the increasing resistance in the parasites to ivermectin.

Our Research into Hollow Oaks

Life in the hollow oaks is in trouble, too. The work we have done in my research group shows that the specialized insects that live in hollow oaks are struggling. In many cases, we have found individual species in very few places, perhaps just a couple of oak trees. These species need areas with many coarse trees that are exposed to the sun—trees that contain a lot of wood mold. There are few such oaks.

Along with other scientists and assistants, I have researched insect life in hollow oaks for more than ten years. We have identified the species of more than 185,000 different beetle individuals from as many as 1,400 unique species in hollow oaks. Some of these are specialists that live solely in oaks or solely in hollow trees, preferably oaks. Around a hundred beetle species that live in hollow trees are endangered or threatened with extinction in Norway.

Today, hollow oak trees enjoy special legal status in Norway:

they are deemed to be a "selected habitat type" precisely because they are associated with such rich diversity. These trees' status as a selected habitat type means we must treat them with special care and avoid damaging them. I am involved in a national monitoring program for hollow oaks, which aims to tell us about their status and development. The hope is to follow this up with a monitoring of the unique insects that live there, too.

If we are to secure these fortresses of biodiversity, we must protect the great hollow oaks we still have left. Our research seems to suggest that traces of intensive oak logging several hundred years ago are still reflected in the diversity of beetles in today's hollow oaks. This may be a kind of delayed reaction, known as extinction debt, whereby species cling on for a long time after habitat destruction but are eventually forced to give up the ghost.

We must also make sure that we prevent the areas around oaks that have developed in an open landscape from becoming overgrown. Many of the most specialized insects do best when the sun can shine on the tree, making it snug and warm. And we must take a long-term view, ensuring that we recruit new oaks that can become hollow well before the old ones die out.

It takes no time at all to fell a hollow oak that's standing in the way of progress—a widened road or a new building. Five minutes with a chain saw, and the giant that put out its first shoots in the days of the Black Death and saw the Renaissance and the Industrial Revolution take shape and pass away, lies splintered on the ground. It takes seven hundred years to replace it with a new oak of the same caliber. And where are the insects to live in the meantime?

7

From Silk to Shellac

Industries of Insects

Throughout history, insects have given us many products of great significance, and many of these products have retained their significance to this day. Some are well known, such as honey and silk. Others you may never have heard of or even realized that they originated from an insect, such as the red coloring in your strawberry jam or the glossy sheen on the skins of supermarket apples.

As always in the case of insects, we're talking about enormous numbers. Even the close to 1.5 billion head of cattle on the planet pale into insignificance when we add up all the livestock in the insect world. According to statistics from the UN's Food and Agriculture Organization, more than 83 billion honeybees buzz around the world in our service. And every year, upward of 100 billion silkworms sacrifice their lives to provide us with silk.

Wings of Wax

Honeybees make honey of course, as discussed in chapter 5 (see page 86). But they also make beeswax, a soft mass produced by special glands in their abdomens, which they use to build nurseries and warehouses for their honey. Beeswax also has many applications for humans and plays a major role in a mythological tale that will be familiar to many.

In Greek mythology, Daedalus and his son, Icarus, flee Crete using wings that Daedalus has created from birds' feathers and beeswax. Daedalus cautions his son against the dangers of complacency and arrogance before they set off: if his son doesn't make enough of an effort, he'll end up flying too low and the sea will destroy his wings; if, on the other hand, he is overcome by hubris and fails to recognize his own limitations, he will fly too high and the sun will melt the beeswax that holds his wings together (a psychologist might perhaps note that the father would have been better off telling his son what he *ought* to do instead of teaching him about all the pathways to catastrophe). Young people clearly didn't listen to their parents in those days, either: Icarus flew too close to the sun, the wax melted, and he crashed into the ocean. But at least he had a sea (the Icarian Sea, which is part of the Aegean) and an island (Icaria) named after him.

Nowadays we use beeswax to make candles and cosmetics rather than wings. The Catholic Church has traditionally been a major consumer because the candles used during Mass had to be made of beeswax. The pale wax was supposed to symbolize Jesus' body, while the wick at the center represented his soul. The flame that burns when the wax candle is lit gives us light, while the wax candle itself burns down—sacrificing itself, as Jesus did, for humanity. Only the very purest wax could be used for this purpose, and bees scored high on that count: since nobody had observed them mating, they were long assumed to be virgins who lived a life of sexual abstinence. Only in the 1700s was this misapprehension corrected (see page 40), but to this day the candles used for Mass in the Catholic Church must contain at least 51 percent beeswax.

It has become increasingly common to use beeswax in cosmetics such as creams and lotions, lip balm, and mustache wax. Honey is also an important component of cosmetics, by the way. If, say, you make one of the many recipes for homemade honey face masks on the internet, you'll be glad to hear that you are in illustrious company: the wife of the Roman emperor Nero, Poppaea—who, of course, didn't have the option of ordering products from the online outlets of the finest French cosmetics companies—made her own face masks from honey mixed with asses' milk. At least that means it doesn't matter if you happen to get some on your lips. Indeed, beeswax mixed with vegetable oils is an excellent lip balm.

Beeswax is also used to help oranges, apples, and melons keep better and to make them look shiny and tempting. This familiar foodstuff is applied to the surface of fruits, nuts, and even food supplement pills. A significant amount of the beeswax extracted from hives these days is also used to produce new beeswax frames that are placed back into the hives. A proper thank-you present!

Silk: A Fabric Fit for a Princess

Silk billows beautifully, is strong but light, is cool against the skin, and has a special sheen all its own. It's an exclusive fabric. It's hardly surprising that in China, silk from silkworms—the larvae of the *Bombyx mori* moth—was long reserved for the emperor and those closest to him.

The history of silk reads like a tale from the Arabian Nights: it's exotic yet brutal, and it's difficult to separate fact from fiction. Two strong women play a central role in the legend. In the beginning, 2,600 years before the Western Common Era, the Chinese princess Lei-tsu was sitting drinking tea under a mulberry tree in the garden of the imperial palace when a silkworm cocoon fell out of the tree and into her cup. Lei-tsu tried to fish it out, but the heat of the liquid dissolved the cocoon, transforming it into the most beautiful thread—long enough to cover the entire garden. In the innermost part of the cocoon lay a tiny larva. Lei-tsu immediately grasped the potential of this discovery and got the emperor's permission to plant more mulberry trees and breed more silkworms. She taught the women at the imperial court to spin the silk into a thread that was strong enough to be woven, thereby laying the foundations of Chinese silk production.

Silk production would remain an important cultural and economic factor in China for several thousand years. Indeed, the country is still the world's largest silk producer, and to this day the cocoons are placed in boiling water to kill the larvae and loosen the thin silk threads.

China guarded the secrets of silk for a long time. Eventually, the trading routes known as the Silk Road opened up between China and the Mediterranean countries, where silk was an important product because the Romans loved it. That said, some viewed this new, almost transparent fabric as immoral; indeed, certain people went so far as to claim that silk dresses were practically an invitation to adultery because they left so little to the imagination.

Be that as it may, we might speculate whether it was actually the amount of gold leaving the Roman Empire to pay for the silk that people found immoral rather than the fabric itself—because China's monopoly on silk production earned the country an enormous income. Consequently people were strictly forbidden to share the secret: an attempt to smuggle out silkworm larvae or eggs was punishable by death.

In the end, the secret came out anyway, and once again a woman played a central role if we are to believe another of the many legends. It is said that a Chinese princess married the prince of Khotan, a Buddhist kingdom in the west of modern-day China that lay along the Silk Road. On her departure, the princess smuggled out silkworm eggs and mulberry tree seeds in her headdress. In that way, the secret spread, the monopoly was broken, and several other countries started to produce silk. Today, more than 200,000 metric tons of silk are produced each year to make clothing, bicycle tires, and surgical thread. Silkworms are still the main producers, although a few other related species are also used.

Hanging by a Thread

Silkworms are not the only insects that spin silk. This skill has probably cropped up more than twenty times among insects over the course of evolution. Green lacewings, for example, fasten their eggs onto small stalks of silk. They look like tiny Q-tips, with the eggs like a clump at the end, and their purpose is to stop ants and other starving souls from getting the eggs. Caddis fly larvae spin silken trapping nets in streams, using them to capture small creatures for their dinner; the larvae of certain relatives of the mosquito, called fungus gnats, spin a trapping net that they use to gather up spores beneath fungi or to trap small insects. Some fungus gnat larvae are even luminous, emitting a blue-green light; nobody can explain why. Unlike the luminous fungus gnat larvae in the caves of New Zealand, which are predators and use the light to lure food into the net, European *Keroplatus* species seem content to obtain their protein from fungus spores and have no obvious reason to play at being light bulbs.

Among some species of dance flies the males use silk to pack up a delightful "nuptial gift" for the female. The males themselves are not predators—they subsist peacefully on a diet of nectar—but they'll do anything for their greedy, protein-crazy inamoratas. So they trap an insect (preferably another male, because that reduces the competition for females—two birds with one stone, so to say) and wrap their prey up beautifully in silk produced by special glands on their forelegs. A suitor bearing gifts that he's even taken the trouble to package himself—it sounds delightful, but the reality isn't especially romantic. This behavior simply shows the hidden hand of evolution at work, as always. One theory is that the bigger the present and the better the packaging, the more time the male gets to mate. Consequently he transmits more sperm and has a greater chance of passing on his genes. And it's fantastic for the female to receive a hefty dose of protein, because laying eggs is an energy-intensive business.

But there's always the odd trickster who'll try to get the benefits without putting in the effort: some males give the female an empty ball of silk—and then have to get the mating over with pretty fast before the lady discovers she's been tricked.

Weaving Miracles: The Spider's Silk

We can't talk about silk without mentioning spiders, even though they are arachnids rather than insects. The group takes its name from the person who became the first spider, according to Greek mythology: a talented weaver called Ariadne, who had the temerity to challenge none other than Athena, the Greek goddess of war and wisdom, claiming to be a better weaver than she. The punishment for her arrogance was to be transformed into a spider. And what an ancestral mother Ariadne turned out to be. Today, we know of more than 45,000 species of spiders. The silk isn't just to make webs to trap their prey; it's also a kind of compensation to the arachnids for lacking the wings of their distant relatives the insects, which they can only envy. By climbing up to an airy spot and producing a long silken thread that the wind can catch, small spiders can sail away on the breeze using their own kiting technique.

Spider silk has impressive qualities. On a per weight basis, it is six times as strong as steel, but at the same time it is highly elastic. This is why a heavy fly that blunders into a web can't simply pass straight through it. Instead, the web gives, a bit like the arresting cables that help fighter planes land on aircraft carriers. This enables thin fabric made of spider silk to halt a flying projectile—a property that can be used to make extremely light bulletproof vests, superabsorbent helmets and a new type of airbag for cars. If only we could learn how to get hold of sufficient amounts of spider silk.

Experiments have shown that it is possible to harvest around 110 yards of silk from a single spider, but it's when you want to scale up that the trouble starts. Unlike the fat, laid-back silkworm larvae, which think of nothing but eating the leaves of the mulberry tree and spinning silk, spiders are predators who have no qualms about eating one another. So it isn't especially easy to keep them in captivity in order to set up industrial-scale silk production.

A beautiful golden silk dress woven from the silk spun by golden orb spiders from Madagascar broke records for visitor numbers when it was exhibited in the Victoria and Albert Museum in London in 2012. That's hardly surprising, because it is a truly remarkable garment that was four years in the making. Every morning, eighty workers collected new spiders. They were hooked up to a small hand-operated machine, where they were "milked" of their silk and then released again in the evening. In all, 1.2 million spiders were needed.

It's plain to see that this is an unsustainable option for industrial production, so people have started to think along different lines. In 2002, the first "spider goats" saw the light of day. With the aid of gene technology, scientists simply transferred "spinning genes" from a spider to a goat, which then began to produce milk that contained the proteins involved in silk production. This sparked considerable media attention but hasn't yet yielded any concrete results to speak of. Norway's neighbors have also thrown themselves into the race to produce synthetic spider silk. The Swedes recently reported that they had produced more than half a mile of thread using water-soluble proteins produced by bacteria. The protein solution solidifies into spider silk when chemical conditions are altered, exactly the same thing that happens at the opening of the spider's spinnerets.

We're still a long way from commercial production, and perhaps that's not so surprising: after all, spiders have had around 400 million years to perfect their silk.

Thank Insects for 700 Years of Notes

Shakespeare's plays and Beethoven's symphonies. Linnaeus's flower sketches and Galileo's drawings of the sun and the moon. Snorre's Sagas and the US Declaration of Independence. What do all these things have in common? They were all written using iron gall ink, a purplish black ink that we have an insect to thank for: the gall wasp. These tiny insects are parasites on plants and trees, and they are most commonly found on oaks. Gall wasps secrete a chemical substance that triggers a growth on the plant that forms a house and pantry around one or several larvae.

There are many varieties of galls. One type that is often used for ink is the oak gall, also known as an oak apple. It does, in fact, look like a small apple—perfectly round with a reddish tinge—except that it happens to be stuck to an oak leaf.

Inside their oak apples, gall wasp larvae lie chomping away on plant tissue in peace and quiet, protected against all foes. Well, only partly, because some parasites have parasites of their own: unwelcome guests who turn up for dinner uninvited and refuse to leave again—such as guest gall wasps, which simply move into other wasps' galls because they can't make their own. Worse still are the interlopers that use their long egg-laying stingers to poke their way through the walls of the gall and lay eggs in the very gall wasp larva that is living there. As a result, the insect that hatches out of a gall may be very different from the species responsible for the gall's creation in the first place.

The walls of the oak gall are stiff with a form of tannic acid. This acid occurs naturally in many plants and trees and is the substance that links your leather jacket with a fine red wine. Tannic acid is crucial for tanning hides and leather, and a master wine connoisseur can distinguish grape varieties and storage methods based on the tannins in a wine.

The first types of ink, made in China several thousand years before the Common Era, used carbon from lamp soot. The soot was mixed with water and gum arabic, a natural gum obtained from acacia trees, which kept the soot suspended in the liquid. But if you were unlucky enough to spill a cup of tea over your writing, your thoughts would be lost forever. Carbon ink was water soluble and easy to wash away—which people also tended to do if they ran short of writing materials and needed to reuse what they had already written on.

Later, people learned to make ink from oak galls mixed with an iron salt and gum arabic. The great advantage of this new ink was that it was nonsoluble: it ate its way into the parchment or paper it was written on. What's more, it wasn't lumpy and was easy to make. From the 1100s to well into the 1800s, oak gall ink was the most commonly used kind in the Western world.

If it hadn't been for the little oak gall wasp, it is far from certain that we would have so many well-preserved and legible documents from the great artists and scientists of the Middle Ages and the Renaissance. If we'd had only lampblack ink, many ancient thoughts, tunes, and texts would have been washed away by water, either because storage conditions were poor or because somebody wanted to reuse the parchment.

Carmine Red: Spaniards' Pride

Insects provide us with colors other than the brown-black hue of oak gall ink. They are also responsible for a beautiful, deep bright red color that was, for several hundred years, exclusively produced in the Spanish colonies and that is still in use today in both food and cosmetics.

Carmine dye is harvested from the females of a particular species of scale insect (*Dactylopius coccus*), peculiar creatures the size of a fingernail that are also known as cochineals. Their natural habitat is in South and Central America, where the females spend their entire life on a single spot, wingless and firmly clamped beneath the protective shield of a prickly pear.

The dye was known to both the Aztecs and Mayans long before the arrival of the Europeans, and they bred a variant that yielded a more intense red color. Since this color was both difficult and expensive to produce in late-medieval Europe, dried cochineal bugs were one of the Spanish colonies' most important wares, valued on a par with silver—because carmine was an intense, powerful red color that withstood sunlight without bleaching. The famous red coats of British soldiers were dyed with carmine, and Rembrandt, among others, used the color in his paintings.

Since the dried insects were small and had no legs and it was before the days of microscopes, Europeans were long uncertain whether the grains of carmine were animal, vegetable, or mineral in origin. The Spaniards kept the secret close to their chests for nearly two hundred years to ensure their monopoly and the vast income the little insect earned them.

Nowadays, carmine comes largely from Peru. The dye is used in many red-colored food and beverage products, such as strawberry jam, yogurt, juice, sauces, and red sweets. You will also find it in various cosmetics, such as lipstick and eye shadow.

Shellac: From Varnish to False Teeth

What do jelly beans, phonograph records, violins, and apples all have in common? A substance extracted from an insect, of course. It's a product with an incredible number of applications, yet you have probably never heard about its origins. We are talking about shellac, a resinlike substance that is produced by the lac bug, a relative of the cochineal bug that gives us carmine. There are heaps of these little creatures on the branches of various tree species in Southeast Asia. According to some sources, the name derives from the Sanskrit word *lakh*, meaning "one hundred thousand," and refers to the enormous numbers of these bugs that can be found in a single place. (A brief diversion: the same source has it that the Norwegian word for salmon, *laks*, has the same linguistic origin for the same reason, because of the large numbers of salmon that gather in the mating season.)

There are several species of lac bugs, but the most common "productive" variety is *Kerria lacca*. Lac bugs are members of the true bug family (see page 28) and spend most of their lives with their snouts stuck into plants—a pretty dull existence. But good heavens, the things this little life has given us humans! One science article went so far as to say that "Lac is one of the most valuable gifts of nature to man."

The tradition of cultivating lac bugs goes back a long way. The insect is mentioned in Hindu documents from 1200 BCE, and Pliny the Elder described it as "amber from India" in writings dated CE 77. But it wasn't until the end of the 1300s that Europeans set their eyes on the product, first as a dye and later as a varnish—in other words, a substance to apply to wood to create a glossy, waterproof surface. Furniture, woodwork, and violins were all traditionally treated with shellac.

But shellac turned out to have many more areas of application. For fifty years, from the end of the 1800s right up until the 1940s, shellac was the main ingredient of phonograph records. It was mixed with ground rock and cotton fiber to produce what Norwegians used to call *steinkaker*, or "stone cakes": brittle, breakable 78 rpm records. The sound reproduction was so-so, but the early record players—or "talking machines," as they were known—were tremendous fun. Bear in mind that radio hadn't yet become common; the world's first-ever wireless radio broadcast wasn't transmitted until 1906 in Brant Rock, Massachusetts, and in Norway test broadcasts didn't start until 1923. So for a long time phonograph records provided the only opportunity to host a "virtual" orchestra or band in your own living room.

Record production was at such high levels in the 1900s that US authorities started to worry, because shellac was also important to the military industry, which used it in detonators and as a waterproof sealant for ammunition, among other applications. So in 1942, the US government ordered the record industry to reduce its shellac consumption by 70 percent.

How do these tiny insects produce a substance with so many varied areas of application: varnish, paint, glazing, jewelry and textile dyes, false teeth and fillings, cosmetics, perfume, electrical insulation, sealant, the glue used to restore dinosaur bones, and a raft of other areas in the food and pharmaceutical industries?

It all starts with thousands of tiny lac bug nymphs settling on a suitable twig. With their sucking mouths, they slurp down plant sap; this undergoes a chemical change inside them and oozes out at the rear as an orange resinlike liquid, which hardens when it comes into contact with air. This forms small, shiny orange "rooftops," which initially cover only the individual bugs but gradually merge into one giant roof that shelters the entire colony and can cover a whole branch. After shedding their skins a few times, adult scale insects hatch out, then mate and lay eggs, well protected by the roof. The adults then die, and the eggs hatch into thousands of new nymphs, which break through the resin roof and set off to find themselves a suitable new branch.

In order to make shellac, the resin coating must be scraped off the branches. It is then crushed and cleaned of insect fragments, after which it is ready for market as small amber-colored flakes or dissolved in alcohol.

Most shellac production occurs in India these days. The good thing is that small-scale farmers in rural villages are the ones doing the job. Some 3 million to 4 million people, many of whom have few other means of earning money, are estimated to earn their bread from keeping lac bugs as livestock. Moreover, the production helps keep species diversity rich in the "grazing areas" of this tiny domestic animal. One reason for this is that little or no pesticide is used, because that would also put these creatures' lives at risk.

Lac Bugs' Skin Care Clinic for Dull Apples

Don't those shiny apples in the fruit aisles of your supermarket look delicious? That's hardly surprising, because they've been for a waxing session at the lac bug's skin care clinic. What happens is this: we humans eliminate the apples' natural wax coating when we wash them after harvesting. And without wax, apples quickly turn into wrinkly, unappetizing fare that few people would want to sell and even fewer would want to buy. So apples have to be waxed again—and that's where shellac comes in, like a sort of antiwrinkle cream.

Many other types of fruit and vegetables also undergo a round of shellacking to ensure that they last longer and look more appealing. The substance is approved for use on citrus fruits, melons, pears, peaches, pineapples, pomegranates, mangos, avocados, papayas, and nuts. In 2013, shellac was also approved to buff up hens' eggs in Norway. The idea is to make the eggs nice and shiny and increase their shelf life.

Shellac also turns up as glazing for various sweets, such as jelly beans, sugar-coated chocolates, lozenges, and the like. This glazing agent also goes by many other names: lacca, lac resin, candy glaze, or confectioner's glaze.

Shellac is used in cosmetics, too: in hair spray and nail varnish and as a binding agent in mascara. It is also used in pills in capsule form, and not just to make the surface shiny: because shellac doesn't dissolve very easily in acid, it can be used to make delayed-release pills—in other words, capsules that dissolve only when they reach the gut.

Once you realize just how many strange places this product pops up in, perhaps it no longer seems so peculiar that somebody should call shellac one of the most valuable gifts of nature to man.

ALFR.
NOBEL

8

Lifesavers, Pioneers, and Nobel Prize Winners

Insights from Insects

Velcro is a genius invention. We use it on shoe straps, jackets, children's mittens, and ski ties. It all started with a Swiss engineer who was out hunting with his dog and got annoyed because the mutt ended up covered in burrs every time they came home. That prompted him to take a closer look at those ingenious seed dispersal mechanisms: small hooks that grab hold of passing animal pelts. Hmmm . . . perhaps it was an idea worth copying. And that's how Velcro came to be.

Engineers and designers are increasingly turning to Nature's solutions for inspiration. Nature has had billions of years to refine its solutions, and evolution has come up with countless smart structures and functions.

When it comes to shrewd solutions, insects make a strong showing because there are so many of them and they are so good at adapting. We can use them as model organisms, as we do with fruit flies. We can make them do things for us that we can't do for ourselves, such as crawl into collapsed buildings or help break down plastic. Perhaps they can provide us with new solutions to the antibiotics crisis, improve mental health among the elderly, or even help make intergalactic travel possible. One thing is for certain: we'll be drawing inspiration from them and imitating them for a long time to come.

Biomimicry: Mother Nature Knows Best

According to the *Oxford English Dictionary*, biomimicry is "the design and production of materials, structures, and systems that are modeled on biological entities and processes." There are numerous examples of biomimicry that started with insects. Dragonflies provide the inspiration for drone technology. Black fire beetles, which have heat sensors on their belly—they lay their eggs in the embers of forest fires—are being studied by the US army and others with a view to developing better heat-seeking sensors.

One key discovery that offers huge potential is that in many cases, insects' color is not the result of pigments but of special structures on the surface that reflect certain wavelengths of light. The result is an intense metallic color that shifts depending on the angle you view it from, as with the bright blue morpho butterfly found in the jungles of South and Central America. Knowledge of structural colors may help us create colors that do not fade, as well as improved solar panels and mobile phone screens and new types of cloth, paint, and cosmetics. And banknotes that can't be forged.

Breathe on Your Banknotes

The beautiful longhorn beetle *Tmesisternus isabellae*, whose only known habitat is a tiny area of Indonesia, changes color according to the atmospheric humidity. When the air is dry, the beetle is gold with dark green stripes. If the humidity increases, its golden coloring shifts to red. Chinese chemical scientists recently copied this trick, applying it to printing technology.

The scientists think their insect-inspired ink can be used to print banknotes that are impossible to forge. If you want to check whether your money is genuine, you can simply breathe on it to see if it changes color. In this way, a unique and rare beetle is helping combat forgery and swindling.

So the only thing left to worry about is keeping your banknotes in a safe, insect-proof place, especially in hot climates, where termites eat anything including even a speck of cellulose—including banknotes. Termites in India have, in fact, chewed their way through a fortune on several occasions. In 2008, they gobbled up all the spare cash an Indian businessman was keeping in the village bank, and in 2011, they chomped their way through piles of rupee bills in a bank vault. The total value was well over $125,000.

Termite Technology Creates Energy-Saving Skyscrapers

Perhaps we can forgive termites for eating a few rupees here and there once we see how much money we can save by copying their architectural solutions. You see, termites have given us some excellent ideas for improving energy efficiency in tall buildings by developing a more natural air-conditioning system.

The enormous termite mounds of Africa can tower several yards above the ground, housing millions of white or pale brown social individuals. Despite the baking heat outside, it is always pleasantly temperate inside the mound. And down in the depths, maybe three feet below the surface, her majesty the Queen of Termites lies in her temperate, oxygen-rich throne room squeezing out eggs at a prodigious rate. All around her, thousands of workers tend the fungus gardens that are like the mound's industrial kitchen (see page 77), where food is prepared for millions. But the fungus is picky and thrives only in temperatures close to 90 degrees Fahrenheit; no more and no less. How do the termites manage to keep the interior temperature constant?

It turns out that an ingenious system of air channels uses temperature oscillations outside the mound over the course of the day and night to create a draft that runs through the construction. This "artificial lung" ensures that cool, oxygen-rich air is drawn down while warmer air, rich in carbon dioxide, is driven out.

Architects copied the termites' ingenious design when they built Eastgate Centre, a large office and shopping mall in Harare, Zimbabwe. Although it is one of the largest malls in Zimbabwe, it doesn't have any regular air-conditioning or heating; instead it uses passive cooling, applying the principles used by termites. As a result, the building uses only 10 percent of the energy that would be consumed by an equivalent-sized building with standard mechanized air-conditioning systems.

From Brown Bananas to Nobel Prizes

You're probably familiar with fruit flies, those sluggish so-and-sos that form a cloud when they fly up off your fruit. Irritating as they may be, these tiny red-eyed insects are, in fact, the winners of no fewer than six Nobel Prizes.

Their family name in Latin is *Drosophila*, "one who loves the morning dew," which sounds a lot more poetic than "fruit fly" and reflects the fact that these insects originally inhabited warm, humid tropical climes. Today, many of the species in the fruit fly family are found throughout the world (with the exception of Antarctica). One common feature of the species that are liable to turn up uninvited in your kitchen is that they thrive on rotting, fermenting organic stuff, such as compost, overripe fruit, or the dregs in the bottom of a beer can. There they lay their eggs and develop at record rates.

Of course, they are pretty annoying. We'd much rather insects left our food alone and stuck to the outdoors life. But these critters are actually more important than you think: the *Drosophila melanogaster* fruit fly is the uncrowned king of the laboratory and has been a crucial component of research and lab experiments for more than a hundred years.

Fruit flies have many great traits that make them particularly suitable for research: they are cheap and easy to keep in laboratories, go through their life span at a supersonic pace, and have oodles of offspring. What's more, we have a good grasp of the species' genetic material or DNA, having fully mapped its genome in 2000. Without wishing to insult anybody, I can reveal that your genes are more akin to those of a fruit fly than you might like. For example, one study that examined a selection of disease-related gene sequences in humans found that 77 percent of them also occurred in fruit flies. It is precisely this similarity that makes fruit fly research such a useful way of understanding various phenomena, even in human beings. The flies have taught us a lot about chromosomes and the way that traits are transmitted. This research earned Thomas Hunt Morgan a Nobel Prize in 1933. Thirteen years later, after being fried with massive doses of radiation, the flies helped Hermann Müller win another Nobel Prize for showing that radiation leads to mutations and causes genetic damage. In 1995, the Nobel Prize in Physiology or Medicine once again went to our wee winged pal along with three scientists whose wide-ranging work showed how genes control development in the early stages of fetal life. In 2004, the prize went to research on the fly's olfactory system, and in 2011, it went to work on the fly's immune defenses. In 2017, the fly won its last Nobel—to date—this time for studies of the built-in clock that controls the circadian rhythm of living organisms. These last prizes are particularly good examples of fly research that is highly transferable to us humans.

Even the thing we find most annoying about the fly—its attraction to stuff that is fermenting and preferably contains alcohol—has turned out to be useful. The research into "alcoholism" in fruit flies is a serious and important business but also involves plenty of human parallels that are sure to pep up the conversation at Oktoberfest—such as the fact that excess alcohol makes male flies clingy and sex mad while simultaneously reducing their chances of successful mating. Or that when male flies lose out on the dating market, they "drown their sorrows" by drinking more than male flies who have managed to mate successfully.

As if that weren't enough, fruit flies continue to increase our knowledge of diseases such as cancer and Parkinson's disease, as well as phenomena such as insomnia and jet lag. So a bit of respect might be in order the next time you catch yourself cursing the tiny flies in your kitchen. As you set up a fruit fly trap, maybe you can at least whisper a little thank-you to one of the most important creatures in biomedical research.

Ants Give Us New Antibiotics

Bacteria are increasingly developing resistance to antibiotics. This is a large and growing problem; according to the World Health Organization, this causes more than 700,000 deaths every year. Knowledge of ecology and evolution is a crucial tool in the battle against antibiotic resistance, and insects are contributing to the solutions.

Ants are an especially interesting subject of study. They live close together in large societies and need good defenses against bacteria and fungi to prevent the death of the entire colony. This is why ants have two special glands on their bodies that produce antibiotics. They smear it over themselves and their sisters using their forelegs, and experiments have shown that this activity increases when fungus spores are present in the nest.

Leaf-cutter ants—the ones that take home leaves, which they chew up and use as a base for cultivating fungi (see page 77)—face extra challenges when it comes to fungal infections: other parasitic fungi sometimes try to establish themselves in the ants' fungus gardens. If successful, they can kill both the fungus crops and the ants themselves. So the ants have developed a powerful defense against such invaders: a collaboration with bacteria that live in special pouches on the ant's body and produce a type of antibiotic that kills the fungal invaders. It is a finely tuned collaboration that has been perfected over millions of years. Studies of this cooperation between ants and bacteria offer us good opportunities to identify effective ways of killing fungi and bacteria. Several discoveries have already been patented, including a fungicidal antibiotic derived from leaf-cutter ants called selvamicin, which is effective against infections by the yeast *Candida albi-*

cans, a fungus many of us have encountered in the form of oral or genital infections.

Larval Therapy

I'm always happy to see clothes or jewelry with insect motifs. It doesn't happen all that often, although a beautiful butterfly or a fluffy bumblebee is sometimes allowed to grace a garment. But flies? Rarely. I carried out a small, extremely unscientific test: an internet search for "butterfly jewelry" in Norwegian resulted in around 1,000 hits. If I switched "butterfly" or "blowfly," I didn't get a single hit.

We think of blowflies as vectors of disease, but these insects can actually heal us by feeding off our infected wounds. It sounds revolting, but this is old news. Genghis Khan was a thirteenth-century Mongolian warlord who founded the empire deemed to be the largest by area in the history of the world, stretching from Korea to Poland. He didn't create this kingdom through diplomacy and negotiations, either, but through brutal and ruthless warfare. Legend has it that Genghis Khan always took a wagon full of maggots into battle with him. They were placed on his soldier's wounds, which made them heal more quickly so that the men could be sent back to the battlefield sooner.

This kind of larval therapy was also used with great success during the Napoleonic Wars, the US Civil War, and the First World War. After we discovered the fantastic properties of antibiotics, larval therapy sank into oblivion. More recently, though, it has returned to the fore, largely because of multi-drug-resistant bacteria.

The larvae of the common green bottle fly, *Lucilia sericata*, are the ones most commonly used for this purpose. This fly can be found outdoors throughout the United States. When used for medical purposes, it is essential for the maggots to be sterile before they are placed on the wound, so they are bred in special laboratories. The maggots are often placed in a kind of coarse-meshed tea bag to ensure that they can't escape but are still able to stick their heads through the mesh to get their job done. And their job involves multitasking. The larvae limit the growth of the bacteria in the wound by producing antibiotic-like substances, and substances that alter the pH value of the wound. They also eat the dead wound tissue. In some cases they have also been found to produce substances that promote the growth of new tissue. They eat only dead tissue and pus and do not touch the living tissue around the wound.

One of the more creative experiments involving blowflies was conducted in the early 1900s by the "Maggot King," an Englishman who believed it was wholesome and healthy for people to inhale the vapors of fly larvae. The man had tuberculosis but was convinced that the maggots he bred as bait for his frequent fishing trips were what kept him alive. And he was keen to share his knowledge with other sufferers. So every summer, the Maggot King had several tons of dead animals sent to him, generally from zoos. He would leave them outdoors until they were full of maggots, which he harvested, transferred to special containers, and then placed indoors in what he called maggotoriums, wooden shacks where patients could sit among the containers of maggots and stinking rotten meat, entertaining themselves with a book, a game of cards, or a friendly chat with the other invalids.

I'm sure few readers, if any, will be surprised to hear that this business idea really stank. People could smell the stench of the

Maggot King's farm for miles around, and his views garnered little scientific support. Although several patients actually testified that their health had improved after spending time among the rotting animals, the inhalation of maggot gases never became a commercial success.

But maybe the future will show that the Maggot King was not entirely on the wrong track. Blowfly larvae can apparently produce gaseous emissions that limit the growth of a nonpathogenic relative of the tuberculosis bacterium that is often used as a test organism. Pending further research, those who use live bait for fishing might as well draw an extra deep breath over their maggot tins, just for the sake of their health.

Crickets as Pets

Insects can also help our mental health. It is common knowledge that keeping pets can improve your happiness and health, and in the East, people have kept insects as pets for thousands of years. In China and Japan, in particular, people have often kept caged crickets, relatives of the grasshopper. The primary attraction was their beautiful song, but in thirteenth-century China, it was also popular to hold cricket fights. Indeed, an annual two-day cricket-fighting championship is held in China to this day. And it is just one of more than a hundred traditional Chinese festivals associated with insects.

It is not an uncommon hobby among Japanese children to catch (or, if they live in a town, buy) large male beetles and arrange fights between them. We're talking about some of the planet's largest beetle species here, with powerful horns or long mandibles, which the males use to fight. In Japan, as in the

United States, bus tours are arranged so that people can see fireflies (which are beetles, not flies) dancing in the night at special locations.

Now insect pets are being tested as a method of geriatric care—in Asia, of course. Because what happens if elderly Koreans have a cage full of crickets placed in their charge?

Nearly a hundred Koreans with a mean age of seventy-one were tested for psychological factors such as depression, anxiety, stress level, sleeping difficulties, and quality of life. After that, they were divided evenly into two groups. While both groups received guidance on healthy living and weekly follow-up telephone calls, only half of the subjects were given a cage that contained five chirping crickets. The species in question was *Teleogryllus mitratus*, a garden cricket that lives in Southeast Asia, whose "song" is considered extremely beautiful and pleasing to the ear.

After two months, all the participants were interviewed and tested again. Almost all the old people liked their crickets, and three-quarters of them felt that caring for the insects had improved their mental health. The test results also showed a slightly positive effect on several of the factors that were measured, in particular a reduced level of depression and an improvement in quality of life.

The good thing about a cricket in a cage is that it is cheap to buy and needs little looking after. The old people don't have to take them out for air, clip their claws, or groom their coats. Yet it can be rewarding for them to watch the cricket shuffle around in its cage and sing, and it needs a bit of food from time to time. In fact, it needs you, which is good to know. Caring for a cricket can be the little bonus that gives daily life some meaning for people who are in poor physical health, can't do much, and spend a lot of time sitting alone.

Biophilia: Love of Nature

Fortunately, it seems as if interest in insects is also growing in the West. Many people have become aware of buzzing bees and chubby bumblebees. People are planting nectar-rich flowers, hanging up insect hotels, and building bumblebee nest boxes in their gardens. Many insect lovers are doing an important job by seeking out and collecting (or photographing) insects from new places. It's like a treasure hunt that offers rewarding experiences of nature while increasing our knowledge of insects.

In several places, particularly in warmer climes, you can find butterfly houses, large areas enclosed in a net where butterflies can fly around freely, being admired and photographed. One Norwegian nature photographer, Kjell Sandved, who worked at the Smithsonian Institution in Washington, DC, became world famous for his butterfly alphabet, beautiful close-ups of butterfly wings displaying letters. The overwintering habitats of the monarch butterfly in Mexico draw tourists from all over the globe, and in 2016 half a million people visited New Zealand to admire the luminous fungus gnat larvae in the ceiling of the Waitomo Caves.

These phenomena highlight an issue that concerned the famous insect scientist Edward O. Wilson: the need we human beings have for a deep and intimate connection to nature and other species. Wilson called it *biophilia*, the love of living things. He thought it was a trait that had developed and been reinforced throughout our evolution because it increased people's chances of survival to be in close contact with nature. If you paid attention to flowers, you could find their fruit a few weeks later. And if you were familiar with the species that could harm you or kill you, your chances of survival rose. Many think that our dislike of snakes and spiders can be traced back to this kind of adaptation.

Nowadays, ever more research confirms how important it is for human health and well-being to be close to nature. Older people survive longer if they live close to a green area, regardless of their socioeconomic status. Students learn better if they can see a patch of green outside their window. Children with personality disorders have fewer symptoms after pursuing activities in nature. It was found that when people moved into public housing were randomly distributed between housing with green areas and housing whose outside areas were paved over, those in housing with green outer areas experienced less violence in the home.

When my children were in elementary school, I used to get to join them on trips to the stream in springtime. Small, skeptical ten-year-olds would watch me use a metal sieve on a long pole to fish up brown mud, which I would tip out onto a white plastic tray on the ground.

"Yuck! You're not going to touch that, are you?" somebody whines. But then the miracle happens: the mud settles, and teeming life emerges. Together we gaze at whirligig beetles with two pairs of eyes, which allow them to see clearly both above and below water, and talk about how the silver bubble on the rear end of another beetle is an air bubble that it's breathing in.

Suddenly there's a battle for the plastic tray and the sieve. Everybody wants to find the strange bugs. Forgotten are the cloth shoes that won't withstand the water; forgotten is their fear of getting mud beneath their fingernails.

Those days have left me with good memories, filled with an intense sense of having contributed to something meaningful.

More than half of the world's population now lives in cities, and the number will only increase. Many lack opportunities to visit wilderness areas or have close encounters with wild animals. Luckily, local wildflower areas and urban green spaces can be excellent examples of the natural world, and you're guaranteed to find insects there.

Cockroaches: Humanity's Best Friend?

New ways of living create new problems and subsequently new opportunities for making use of insects. Rescue work in cities, for example in a collapsed building, presents particular challenges. Saint Bernard dogs with casks beneath their chins can't help us here. In the modern-day urban environment your guardian angel may soon turn out to be a cockroach.

You've probably heard the saying that cockroaches are the only things that will survive a nuclear war. It's a myth spawned by old films with thrilling titles such as *Them*, *Bug*, or *Twilight of the Cockroaches*; films dominated by postapocalyptic monster insects that eat atomic fallout for breakfast and any surviving women—preferably young beauties—for dessert. It's nonsense, of course, although it is true that cockroaches can withstand more radioactivity than us humans (by the way, mealworm beetles can withstand even more).

Cockroaches' resilience, not to mention their robust physique and spectacular motor skills, can actually be useful to us. Pack a tiny cockroach backpack full of modern technology: a microchip, a transmitter and receiver, and a control unit that is connected to the cockroach's antennae and cerci, the tactile, taillike appendages on its rear end. The microcontroller, which is operated remotely, can stimulate the cerci with small electric impulses. This makes the cockroach think that something is approaching it from behind, so it runs away. If you send an impulse to an antenna, the cockroach thinks it has touched something and nimbly turns aside. In this way, you can remotely steer a whole armada of backpack-bearing cockroaches through a dangerous building and, by interpreting the signals that are sent back, can draw a map of the accident site.

The backpack can also be supplemented with a microphone that captures audio in the surrounding areas. In this way, the people remotely controlling the cockroach can listen for sounds from people who are trapped, after an earthquake, say. By steering the cockroach toward the sound, they can identify the position and come to the rescue more speedily.

So if you should be unlucky enough to get trapped in a collapsed building, don't be too quick to stamp on any cockroaches that happen to come your way. They may prove to be your salvation. If, on the other hand, you should get lost in the Swiss Alps on a winter's day, you'd be better off pinning your hopes on a Saint Bernard. Snowy weather is one of the few things a cockroach can't deal with.

Plastic on the Menu

Every minute enough plastic is dumped into the world's oceans to fill an entire dump truck. At least as much again ends up in landfill sites, and the amounts are constantly increasing. Because we love plastic. It's handy and cheap. We produce and use twenty times as much plastic every year now as we did fifty years ago, and less than 10 percent of it is recycled. The rest of the plastic waste ends up in landfills, in roadside ditches, or in the sea. A report issued by the Ellen MacArthur Foundation estimated that if this continues the sea will contain more plastic than fish by 2050. This is because plastic biodegrades extremely slowly in the natural environment. So the discovery that a number of insects can digest and break down plastic is something of a sensation.

Take polystyrene, for example. Even if you don't think you use it often, I'm guessing that you've held some in your hand—if you've ever bought takeout food in a carton or a hot drink in anything other than a paper cup. Because polystyrene, also known as isopore, is the material used to make disposable containers for hot food and drink. In the United States alone, 2.5 billion such cups are thrown away every year—and we're talking about a material that was thought to be nonbiodegradable. Until now. Because it turns out that mealworms consume isopore cups as if they were part of their regular diet.

In one study, several hundred American and Chinese mealworms were served some isopore. All of them belonged to the darkling beetle species (*Tenebrio molitor*), which lives outdoors in most parts of the world and sometimes turns up indoors, too, if any soggy flour residue is left lying in your cupboards for too long. They gobbled up the isopore at record speed, and the larvae raised on this peculiar diet pupated and hatched into adult beetles as normal. Within a month, for example, five hundred Chinese mealworms had gobbled up a third of the 5.8 grams of isopore served up to them. All that was left was some carbon dioxide and a spot of beetle poo, which was apparently pure enough to use as planting soil. There was no difference between the survival rates of larvae that received normal food and those on the isopore diet.

But this can hardly be called a superfood. So another experiment compared three different groups: larvae that received isopore, larvae that received some kind of cornflakes, and larvae that received no food at all. The weight of the cornflake-fed larvae increased by 36 percent, while the isopore-fed larvae didn't put on any weight. But they still did better than the poor starving larvae, who lost a quarter of their weight over the two-week duration of the experiment.

Strictly speaking, the beetles themselves aren't the ones doing the job of breaking down the plastic. For this they rely on some tenants in their gut. If the mealworms are given antibiotics that kill off this gut flora, their ability to break down plastic also vanishes. The breakdown of plastic probably depends on the combined efforts of beetle and bacteria.

More research needs to be done into whether this insect can help us solve the problem of plastic in the oceans because mealworm beetles aren't keen on getting their feet wet and are hardly suited to the seagoing life. But there is plenty of plastic on dry land that we'd love to get rid of, too, and these beetles may be able to help.

Mealworms are not alone. Other insects can also help solve the plastic problem. The greater wax moth is a lepidopteran that is viewed as a pest by beekeepers because it eats the wax combs inside beehives. But beeswax has a structure similar to that of polyethylene, which is the kind of plastic used in supermarket shopping bags. And sure enough, it turns out that the wax moth larva can eat holes in this kind of plastic and transform it into ethylene glycol, a substance we know as antifreeze. Again, this task is not performed by the larva alone but is probably a result of bacteria living inside its gut.

Researchers are now poring over these findings to discover how we can mass-produce the active substances and perhaps, over the long term, translate this into a practical solution to help us deal with our plastic waste.

Forever Young: The Beetle with the Elixir of Youth

Sometimes scientific discoveries come about quite by chance—as when a US scientist happened to forget some larvae in a drawer around the end of the First World War. It's probably easy to get into a muddle when, like that chap, you're studying everything from human cell structures and causes of sterility in mules to the caddis fly's reaction to light. But quite *why* that scientist happened to leave a can of beetle larvae in his office drawer in the first place is a mystery.

At any rate, the point is not that he left them there but that he forgot about them. Totally and absolutely—for five whole months. And for the colored cabinet beetle *Trogoderma glabrum*, whose normal life cycle is a mere two months from egg to dead adult, five months without food should have been the end. But when the scientist finally rediscovered the larvae in his drawer, he found them in the best of health. Stranger still, they had grown younger! Yes, they really had!

If you cast your mind back to chapter 1, with its crash course in insect lore, you may remember that all insects shed their skin a number of times en route to adulthood. This is normally meant to go one way and one way only: from tiny larva to larger and more developed larva—just as we humans can go only from baby to teenager and not the reverse. But the larvae in the drawer had, in fact, gone the opposite way: they had developed backward, from big to little, from an advanced to a simpler larval stage.

This was revolutionary stuff. The scientist grasped that much. He continued to starve the beetle larvae and discovered that the insects could stay alive for more than *five years* "without a particle to eat," as he wrote. They simply got smaller and smaller because they were *living their lives backward*—going from the late larval stages to the earliest. Stranger yet, when those poor wretches on enforced hunger strike were given access to food once more, the switch flipped back to normal mode and the development from "baby" to "youth" resumed.

A more recent study from the 1970s confirms those old findings. The larvae of the *Trogoderma glabrum* can develop forward and backward repeatedly. True, the process isn't entirely without cost, because although it may look like a "baby larva," a larva that has undergone repeated back-and-forth cycles will display physiological deterioration, indicating that it has aged after all. With each new round, it takes longer for the larva to grow bigger again.

It's a pretty wild business. And there's more where that comes from: the aging process can also be controlled in honeybees. Bees that are responsible for looking after the young in the hive can live and remain at the height of their mental powers for many weeks. However, worker bees, the ones that go out and gather nectar, die, thoroughly senile, after a couple of weeks. The ingenious thing is that if worker bees are forced to take on the hive bees' job again, some of them actually "grow younger"—they have a longer life with high mental capacity. In honeybees, this is controlled by a special protein, a kind of bee elixir of youth. Studies of these insects can help us understand aging processes in humans, too, which may lead to new insights in areas such as dementia-related diseases, ultimately helping us achieve better health in our old age.

Midges in Space

Speaking of life expectancy and aging, how about a trick that could help us with interstellar travel? Perhaps insects can help there, too. A nonbiting midge known as the sleeping chironomid, *Polypedilum vanderplanki*, is actually a hard-core aspiring astronaut equipped for prolonged periods of sleep.

The midge lives in Africa, and the larvae spends their life in small puddles of water that are constantly drying out. But whereas we humans die if we lose more than 14 percent of the water in our bodies and most other organisms can withstand a maximum water loss of 50 percent, this chironomid can cope with a loss of up to 97 percent! In this desiccated state, the larvae can handle pretty much any pummeling: you can boil them, dip them into liquid nitrogen, soak them in alcohol, expose them to cosmic radiation for years, or simply leave them alone; the record survival time to date is seventeen years.

When it is time for them to wake up, all you have to do is pour water on them, and—Hey, presto!—like the freeze-dried chunks of meat in packages of dehydrated soup, the larvae swell up to their normal size. Give them an hour, and they will be busy eating again as if nothing had happened.

So the midge larva can enter into a state that is somewhere between life and death without suffering any apparent harm. All it needs is a little time to prepare itself. The key to its survival seems to be for it to replace the water in its body with a kind of sugar called trehalose. This sugar is only about half as sweet as normal sugar and occurs naturally in small concentrations in insect blood. Incidentally, trehalose is named after the cocoon-shaped larval secretions of a true weevil (or snout beetle) found in Iran, which is known as *trehala* in Persian and is widely used in Persian traditional medicine.

When the larva realizes that tough times lie ahead, its body begins to produce more trehalose sugar, which rises from the normal level of around 1 percent of its blood up to around 20 percent. The sugar protects the cells and bodily functions in various ways.

Several other organisms have mastered the art of becoming the living dead, including bacteria, fungi (think of dried yeast!), roundworms, tardigrades, and springtails. The exciting thing is that they don't all use the same techniques. In tardigrades, for example, there is no sign of trehalose accumulation.

If we can get to the bottom of the processes that control this switch between normal life and desiccated dormancy, we can use it to keep cells, tissue, or perhaps even individuals in a desiccated state. Perhaps an African midge will help us find the key to interstellar travel in the future!

Robot Bees

While we wait for insects to help us travel to the stars, maybe we can appreciate the fact that they travel among the flowers—simultaneously pollinating them. There's plenty of inspiration to be found here. Robot bees actually do exist, in laboratories at any rate: tiny drones that have been enhanced with a brush and electrically charged gel so that they can gather pollen. Carbon-fiber brushes, nylon hair from a makeup brush (yes, really), and horsehair have all been tested, and although horses aren't known for their pollinating skills, the horsehair brush worked best. With that, robot bee 1.0 was ready for testing. You can find a video clip on the internet of the drone flying from lily to lily in the Japanese laboratory where it was created. The flight itself is rather clumsy, but then again, drone flying isn't on the university curriculum—yet.

The most obvious area of application for these sorts of drones is on pollination-dependent food crops in greenhouses. That might allow us to restrict the use of introduced bumblebee species, which have a habit of escaping from the greenhouses and dispersing in the natural world. For now, the robot bees aren't especially efficient because they have to be controlled manually and need constant recharging, but perhaps in the future they will be able to navigate using GPS or be controlled by artificial intelligence and will have batteries with a longer life.

Let's hope we don't end up in a world in which we believe modern mechanics can replace nature's infinitely advanced functions. In nature, more than 20,000 different species contribute to the pollination of wildflowers and crops, and research shows that pollination is most effective when it involves a diverse range of species with different special adaptations. We know that the interplay between insects and flowers has been fine tuned over more than a hundred million years and that natural pollination is far more complex and ingenious than any imitation we might come up with. It is simply easier and cheaper to conserve the solutions that nature gives us free of charge.

When it comes to gaining new insights from old insects, we never know which species will turn out to be useful next. Mealworms, fruit flies, or cockroaches; ants or midges. We humans are quick to categorize other species according to whether they are a help or hindrance to us. And we're generally anxious to get rid of the ones that fall into the latter group. But nature is so cleverly organized that, with better knowledge, we will always be discovering new smart solutions. This is one reason it is so important to conserve the natural world and all the species that live in it, whether we consider them useful or not.

9

Insects and Us

What's Next?

The extraordinary lives of insects are changing. The ecosystems on Earth have altered more rapidly in the past hundred years than at any previous time in human history. Well over half of the planet's land area has been transformed through agriculture, livestock grazing, and construction. And the pace is increasing. This means that habitats are being lost and those that remain are being divided into small, isolated fragments. Dams and artificial irrigation are putting the planet's freshwater resources under ever-increasing pressure. We have produced and thrown away so much plastic that we will be finding vestiges of it in the sediments in the form of microplastic for generations to come. Every year we produce substantial amounts of chemicals, including the pesticides we use to protect our crops, which kill insects. We displace species, both intentionally and unintentionally. We have doubled the amount of nitrogen and phosphorus in the soil through the use of artificial fertilizer, and carbon dioxide emissions are higher than they have been for tens of millions of years, contributing to climate change.

All of this affects insects. And anything that affects insects affects us. A decline in the number of insects and the extinction of species will spread through the ecosystem in a ripple effect with major consequences over time because of the impact on so many fundamental ecological functions.

Fortunately, we'll never manage to wipe out all the bugs. But we would do well to take more care of our tiny six-legged winged friends, because despite their 479-million-year track record, they're starting to struggle.

We know of only a tiny proportion of all the insect species in existence, and we have little solid monitoring data about those we do know of. Even so, one estimate indicates that one-quarter of all insects may be under threat of extinction.

One important point in this connection: it's too late to worry when a species is on the brink of extinction. Species cease to function in the ecosystem long before the last individual dies out. That is why it is so vital not to focus exclusively on species extinction, but also to turn the spotlight on the decline in the number of individuals—and there is much to suggest that insects are becoming fewer. A German study in 2017 found that the accumulated biomass of all insects trapped in more than sixty locations nationwide had plummeted by 75 percent in just thirty years. These trends were confirmed by a German study from 2019, using different methods but still finding the same trends.

Still, it is important to realize that the pattern of change is complex. The small amount of data we have indicates that not all insect species decline, and not in all places, but the lack of standardized, long-term datasets makes it challenging to quantify insect decline on a global scale.

Why are insect numbers falling? It isn't easy to say, because there are probably many connected causes. Important factors include increasing land use, intensive farming and forestry practices, pesticides, and the decline in natural remnant habitats, as well as climate change.

What happens when our demands for constant growth in the use of land and resources cause insect populations to crash, species to disappear, and insect communities to change? Think of the world as a hammock made of woven fabric: all the species on the planet and their lives form part of the weaving, and all of them combined create the hammock we humans are resting in. Insects are so numerous that they account for a large share of the hammock's fabric. If we reduce insect populations and wipe out insect species, it's as if we are pulling threads out of the weave. That might be fine as long as there are only a few small holes and loose threads here and there. But if we pull out too many, the whole hammock will eventually unravel—and our welfare and well-being along with it.

Drastic changes in insect communities can have domino effects with consequences nobody can predict. Indeed, we do not know what their significance might be—only that things may become very different. We risk living in a world where we humans face a tougher existence because the challenges of ensuring clean water, sufficient food, and good health for all become even greater than they are today.

Let us look at a few challenges, some examples of the factors that threaten insect life, both locally and globally.

First: land use. This is undoubtedly the greatest threat. We are using land ever more intensively. That means fewer habitats, fewer intact rain forests in the tropics, and, back home, fewer flower meadows in agricultural land and densely built-up areas and fewer areas of natural forest, where old dead trees get to play out their role as homes for insect diversity. It also means more artificial light, which has an impact on many insects.

Second: climate change. Warmer, wetter, wilder—that's the outlook. What do these changes mean for insect life?

Third: challenges related to the use of pesticides and new genetic manipulation techniques. This is a vast field that leaves us with more questions than we have answers for.

Fourth and finally: the introduction of nonnative species and their effect on bugs. What is the right way of dealing with "past sins" in this area? Is it possible to reverse them, and is this the correct prioritization? Because at the same time as we wipe out species, the changes we make create room on the planet for new species, brought to the fore by the thrust of evolution. How robust is nature, and how shall we weigh our concern for our own kind against our concern for millions of other species?

The Frog You Wouldn't Want to Kiss

In the South American jungle lives a horribly poisonous frog with the appropriate Latin name of *Phyllobates terribilis*. In English it goes by the name golden poison frog. This isn't a frog you'd want to kiss in the hope of finding a handsome prince. If you tried, you'd be dead within minutes—guaranteed. The poison in question is one of the most powerful nerve poisons known, batrachotoxin. An average frog contains roughly one milligram of the poison, which is about the same weight as a grain of salt. This alone is enough to kill ten grown men. And just so you know: there's no antidote.

This little frog, no larger than a plum, used to be fairly common in the rain forest in parts of Colombia. The locals would carefully stroke their arrows along the frog's back to ensure that their arrowheads were poisonous enough to kill anything they might encounter.

The pharmaceutical industry got wind of this shocking yellow poisonous sensation in the rain forest. Early tests indicated that the poison was an incredibly effective painkiller—in suitable doses. What's more, because it affects the transport of sodium through cell membranes, it could also be significant for our understanding of numerous diseases where this is important, such as multiple sclerosis. A few specimens were fetched from the jungle for closer examination. But guess what happened to the catch when it arrived in the laboratory? The frog was no longer poisonous!

The fact is that nature is often craftier than we expect: the golden poison frog is not poisonous in itself. It produces the poison only while living in its natural habitat. Why? After much laborious detective work, we now know that the poison comes from a diet of beetles—yes, that's right (this is a book about insects after all). A beetle from the soft-winged flower beetle family Melyridae, to be precise. So the frog is poisonous only when it eats the right kind of beetles in its natural forest habitat.

Thanks to rain forest logging, the golden poison frog is now listed as threatened with extinction. A desperate battle to rescue the species is under way, but there are few bright spots. Not only is the frog's habitat vanishing, but the trade in frog legs has led to the spread of a fungal disease (popularly known as Bd) that is killing frogs, toads, and newts around the globe. A third of them are now on the point of vanishing for good. Soon there will no longer be any golden poison frogs and thus no opportunity to do further research into the active ingredients they produce.

Varied Landscapes Increase Insect Numbers

If we want to preserve our chances of seeking out active medical ingredients, we need to take care of the habitats of these species. Conserving natural areas intact is one important means of securing habitats, both in the rain forest and in industrialized nations. Many specialized insects have such peculiar special needs when it comes to where they can live that they cannot survive in the totally transformed modern landscape. This means that nature reserves and other conservation areas are crucial if we are to safeguard unique species.

But it is also important to retain as much of the variation seen in the natural landscape in places outside the large conservation areas. In the forest, that may mean ensuring that there are enough old and dead trees—because dead wood plays a central role in a living forest (see page 112), housing a large share of forest species, including insects, which make themselves useful as decomposers, pollinators, dispersers of seeds, food for other animals, and pest controllers. And although many countries have recently launched initiatives to increase the amount of dead wood, the volumes are still low compared to natural conditions.

In our farmland and cities we can also achieve a great deal through simple measures that simultaneously serve to beautify the environment for us bipeds: a belt of trees and bushes alongside a stream in a residential area; green shoulders and hedges along roads, and a border of wildflower meadow along the edge of a field; a patch of uncultivated land in the middle of a field with old hollow oak trees. A varied landscape provides many more opportunities for complex insect life. Again, this benefits the pollination of both wildflowers and crop plants. Honeybees, wild bees, and bumblebees aren't the only insects required for good and efficient pollination; this is high-level teamwork involving many players. It is often the case that flies, beetles, ants, wasps, and butterflies are less efficient pollinators *per flower visit* than bees and bumblebees. Yet this is often offset by the fact that they visit many more flowers overall because there are such a lot of them. Some of these "nonbees" may also have peculiar habits and adaptations that benefit pollination.

If we combine the data from a few dozen research projects from five continents about crops of, say, rapeseed, watermelon, mango, strawberries, and apples, we find that the plants produce better crops (an increased fruit set) when they receive visits from "nonbees" regardless of how many bees paid them a visit. It seems that these other insects contribute something unique, something that bees cannot deliver. There are also differences in how vulnerable the different insects are to changes in the landscape, which is an advantage for our food production. In sum, all these insects operate as a kind of pollinator insurance: if one species can't get the job done, another one can step in.

We know that species diversity can make the ecosystem more effective when it comes to capturing resources, such as water and nutrients, and that this results in more biomass. That knowledge is key once we grasp that this biomass is precisely what serves as the basis of crops and the food that ends up on our dinner table. We also know that species diversity is central to breaking down the biomass once again and thereby ensuring that the nutrients are released, which enables new production.

What's more, we are gaining increasing support for the notion that intact biological diversity can make ecosystems more stable over time than impoverished diversity does. There are many mechanisms in play, including the fact that different species have different strengths: one species grows best during cool summers, while another does so beneath a baking summer sun. When species decline or are wiped out, nature has fewer variants to play with and we are more poorly armed against both natural fluctuations and man-made changes—in the climate, for example.

It isn't easy to put a price tag on the services insects provide, although that hasn't stopped people from trying. For example, the annual global contribution of the many pollinating insects is estimated to be worth around $577 billion. Decomposition and soil formation are estimated to be worth four times as much as pollination in total. Although these figures vary depending on calculation methods and are pretty approximate, they still show that the contribution by insects is extremely valuable and that it makes good economic sense to encourage it.

Troublesome Light

The fact that we humans are spreading out over increasingly larger expanses of the planet also has some consequences we don't think about on a day-to-day basis, such as light pollution, the sum of artificial outdoor light produced by streetlamps, houses, and industrial buildings. Light pollution is growing at a rate of 6 percent per year and is disturbing our ecosystems, including insects.

We all know that moths are attracted by light, although the exact cause is a matter of debate. According to the leading theory, they think the light is the moon and try to orient themselves by maintaining a fixed angle to it. While this works perfectly well with the moon, which is a long way off, the result in this case is that they spiral in toward the artificial light and generally end up getting fried.

Street lighting can alter the local species composition of bugs. When artificial light is reflected off shiny surfaces, this can confuse land-living insects that lay their eggs in water. Where we see a parked car beneath a streetlight, a dragonfly perceives the light as being reflected off the surface of a body of water and drops its entire life's egg production in the wrong place.

What will happen to insects over the long term? Could light pollution cause urban insects to change their behavior and avoid light, for example? To test this, some Swiss scientists compared a thousand larvae of the spindle ermine moth species (*Yponomeuta cagnagella*), half of which came from the city and half from the countryside. All got to spend their childhood in similar lighting conditions in a laboratory. Right after hatching, as night fell, they were let out into a large net cage, with a light source placed on the opposite side. Then it was just a matter of waiting the whole night through. Would city moths and country moths be equally attracted to the light?

The result was clear: city moths were evidently much less attracted to the light, by an average of as much as 30 percent. This indicates that nocturnal moths that have spent generation after generation living in an artificially lit environment have undergone an evolutionary adaptation to artificial light. After all, it doesn't make much sense for masses of them to fly round and round streetlamps getting burnt or eaten up by predators that have worked out where the buffet is being served. This could explain the emergence of selective pressure against attraction to light among urban moths.

On the one hand, this is fine, because it prevents them from dying like that. On the other, it can have far-reaching adverse consequences. Because there is a cost attached: the avoidance of light probably means that urban moths simply spend more time sitting still.

Consequently, the effect of artificial light in built-up areas alters the insects' role in the ecosystem. For example, it is difficult for nocturnal insect eaters to catch an insect that is hidden and motionless. Nor is an insect that can't be bothered to fly going to do much pollination among flowers that are adapted to nocturnally active pollinators. That is why it is important to restrict light pollution and, in particular, to try to keep artificial light away from natural areas that are not yet affected by it.

Warmer, Wetter, Wilder—What About the Beetles?

We know that we are heading toward a future in which the climate will be different. This will affect insects both directly and indirectly.

One challenge is that climate changes disturb the finely tuned synchronizations between different species. We see a shift in the timing of many processes, such as the return of migratory birds and foliation, or spring blossoming. The challenge is that different events do not necessarily shift in sync. If insect-eating birds produce their young too late or too early in relation to the period when there are the most insects, there may be too little food for the chicks in the nest. This can happen if some events are triggered by length of day (which is not affected by global warming) while others are triggered by mean temperature (which is), for instance. In the same way, plants that are reliant on particular insects for pollination may suffer from poor seed production if they flower at a point when these insects are no longer swarming.

The spring can be particularly challenging, especially a "false spring" that arrives far too early. When that happens, overwintering adult insects are tempted by the warmth to go out in search of food. When the frost returns, the insects will struggle to cope with the cold and with finding enough food because they have poor cold tolerance and few food reserves.

We see that many insects try to change in response to changes in the climate. Sometimes their entire distribution is shifted, but we often see that the species fail to keep up and the distribution shrinks instead. In the case of dragonflies and butterflies, it has been proven that many species have become less widespread and are shifting northward. Color charts of the different insect species show that many butterflies and dragonflies, especially those with dark coloring, have vanished from southern Europe and sought refuge in the northeast, where the climate is cooler. Scenarios produced for bumblebees indicate that we may risk losing between a tenth and—in a worst-case scenario—half of our sixty-nine European varieties by 2100 owing to climate change.

In the north, climate change is increasing the distribution of leaf-eating caterpillars. This exacerbates the effects on the birch forests, which are being chewed bare. Over the course of a decade, outbreaks of autumnal moths and their relatives have caused considerable damage to the birch forests of Finnmark in northern Norway. The outbreaks have ripple effects on the entire system: food conditions, vegetation, and animal life are all changed.

Along with researchers in Tromsø and at the Norwegian University of Life Sciences outside Oslo, I have looked at how the autumnal moths' depredations affect a different group of insects: the beetles that break down the dead birches, thereby ensuring that the nutrients are recycled. Our results show that the attack of the autumnal moths creates so many dead birch trees in such a short space of time that the wood-living beetles are simply unable to keep pace. They cannot respond to the increase in available food with an equivalent increase in the number of individuals. We do not know what effect this may have over the long term, and this illustrates a key point: we have no idea what sort of consequences continued temperature increases will have for the ecosystem in the north, but it is obvious that there will be dramatic changes.

Since one of my research fields is insects in large, ancient, hollow oak trees, I have been wondering how climate change will affect the beetles that inhabit them. A couple of years ago, my research group and some Swedish scientists compared a large data set that covered beetle communities associated with oak trees across the whole of southern Sweden and southern Norway. The oaks stood in places with differing climates, so that the range they spanned in terms of temperature and precipitation was roughly equivalent to the changes foreseen in climate scenarios. We looked at differences in the beetle communities in order to gain knowledge about how a warmer, wetter, and wilder climate might affect these different insect communities in the future.

In our study, we found that warmer climates were good for the most specialized and peculiar species. Unfortunately, though, these unique species reacted badly to increased precipitation. This means that climate change is hardly going to improve conditions for these particular insects. However, the more common species showed few reactions to climate differences.

This confirms a pattern that is common in our times, not just in relation to climate change but quite generally: locally unique, specially adapted species are the ones that suffer, whereas common species do fine. This means that many rare and unique species will go into decline, whereas relatively few species that are already common will become more common. This is known as *ecological homogenization*: the same species are found everywhere, and nature becomes more similar across different geographies.

Insecticides and Genetic Manipulation: Dare We—Should We?

Every year, we use massive amounts of chemicals quite intentionally to kill insects. After all, that's the whole point of insecticides used in agriculture and private homes and gardens.

Many people think the intensive use of pesticides in agriculture is a price we have to pay to be able to feed a constantly increasing population through industrial agriculture. Others argue that we ought to take a more ecological approach and cooperate with nature in our agricultural practices, even if it might lead to lower crop yields.

Although we don't have space to go into this discussion here, I must mention the large and growing documentation detailing the unwanted harmful effects of neonicotinoids, a widely used group of insecticides. These substances affect the navigation and immune defenses of honeybees and wild bees and may be among the reasons for the decline in these groups.

We humans have recently acquired a brand-new tool in our battle against insects that are harmful to us. I'm thinking of genetic manipulation, in particular what is somewhat cryptically known as the CRISPR/Cas9 method. This is like a pair of molecular scissors that can cut up genes and may be used to alter an organism's DNA by removing or substituting certain genes. The method may be combined with something called a gene driver, which ensures that the genetic change rapidly spreads to pretty much all offspring.

Malaria is caused by a parasite that a mosquito carries from one infected person to another when it sucks their blood. Every year, around half a million people, most of whom are under five years of age, die of malaria. Even so, the numbers are much lower than they were just fifteen years ago, and much of this decrease is due to simple measures, such as the use of insect nets impregnated with insecticide. But now we also have a tool that could ultimately be used to wipe out the malaria mosquito once and for all. This can be achieved by making one of the sexes sterile or by ensuring that all the offspring are the same sex.

This has prompted a timely question from the Norwegian Biotechnology Advisory Board in several different forums: Dare we—and should we—use such tools in the natural world? We know little about their impact. One issue is that we don't know what cascading effects doing so could have in the ecosystem.

What if we eliminate one species and another simply steps in and takes over as a spreader of disease? For all we know, things could end up worse than they were to start off with.

Another question is whether the use of a tool like this could lead to undesirable mutations, with undreamed-of consequences. Terrifying scenarios, such as the spread of sterility to other organisms, lie in wait. Despite the need for haste, we must proceed with caution. Before we start to use the new gene technology tools to genetically alter or wipe out insects that spread serious diseases, we need to protect ourselves against undesirable consequences as best we can.

The End of the Giant Bumblebees

We humans have changed a lot of things on this planet. Some are things we cannot change—such as the fact that our forefathers wiped out most of the really big animals tens of thousands of years ago on continent after continent. Gone are the mammoth, the saber-toothed tiger, and the giant sloth. Along with them, a great many insects probably died out, too, insects associated with this megafauna in various ways, although we know even less about them.

Other changes are much more recent. Seagoing explorers took cats, rats, and other efficient predatory mammals to islands where life had gone its own way. Native species that hadn't the wits to look after themselves, species that had developed in the absence of such enemies, were then often summarily dispatched.

We continue to displace species at a rapid pace, sometimes without meaning to and sometimes intentionally. One example is the import of buff-tailed bumblebees to South America, where they were supposed to improve pollination in fruit orchards and greenhouses. The buff-tailed bumblebee has spread rapidly and is crowding out the local giant bumblebee, *Bombus dahlbomii*, apparently because the buff-tailed bumblebee carries parasites that the giant bumblebee can't cope with. The *Bombus dahlbomii* is the world's biggest bumblebee, affectionately described by bumblebee expert Dave Goulson as "a monstrous fluffy ginger beast." Soon it might be gone forever.

What are we to do with introduced species that threaten unique native species? This is a big, difficult, and important question that we need to debate more in society. In some cases the decision forces itself upon us, as in New Zealand. The government there has launched a plan to wipe out rats, opossums, and stoats. These alien species kill roughly 25 million birds every year.

Many other island nations suffer from the same problem. The challenge can be illustrated by a tale from Australia about a once extinct stick insect that was rediscovered and the living, but soon-to-be dead, black rats that ate it up.

Routing the Rats?

On June 15, 1918, the steamship SS *Makambo*, packed to the gunwales with fruit and vegetables, ran around just off Lord Howe Island, a tropical island far out in the Pacific Ocean that was an easterly outpost of Australia. Its few inhabitants were separated from the mainland by more than 360 miles. The important point about this shipwreck is the creatures that managed to reach dry land: rats. In the nine days it took to repair the ship, an unknown number of black rats managed to reach the shore and establish a foothold on the island.

Lord Howe Island had lain isolated in the middle of the ocean for millions of years. Unique species had developed there that existed nowhere else on Earth. But the rats hadn't come to chill on the beach. Remember the story of the very hungry caterpillar? The one that ate its way through an apple on Monday, two pears on Tuesday, and ended up plowing its way through oranges, sausages, ice cream, and chocolate cake before the weekend was over? That's roughly what the rats did on Lord Howe, too, the only difference being that they ate up unique species one by one. In the early years alone, they finished off at least five species of birds and thirteen small animals that couldn't be found anywhere else in the world.

One of these small creatures was a gigantic stick insect—you know, those thin pale brown insects that look like dried twigs. But this species wasn't just any old stick insect. We're talking about a quite special insect, the world's heaviest stick insect: it was the size of a big barbecue sausage, dark, shiny, and wingless, and it aptly went by the nickname "tree lobster." Its Latin name, should you wish to know it, was *Dryococelus australis*. It proved to be a delicious meal for hungry rats. As early as 1920, the species was already declared extinct—a kind of belated victim of the shipwreck two years earlier.

But this story takes an unexpected turn. Because the outpost has an outpost: 12 miles away from Lord Howe Island lies Ball's Pyramid, a sheer narrow sea stack that is taller than the Empire State Building. For years, it attracted adventurous climbers, but since being awarded World Heritage status in 1982 (along with

Lord Howe Island), only scientific expeditions have been allowed to visit. Around that time rumors also began to circulate that there were "tree lobsters" on the sea stack. Suddenly there was no end to the number of climbing expeditions with an inordinate interest in bugs that were applying for climbing permits to go looking for this rare stick insect. In the end, the man in charge got so sick of issuing climbing permits disguised as insect research that he decided to put an end to the rumors once and for all.

Thus it was that, in 2001, two scientists and two assistants traveled to the sea stack. They climbed up the sheer rock face without seeing a single tree lobster, but on their way down they discovered a little bush of the type eaten by the insect wedged into a crevice in the rock face. Below it lay some large droppings, which looked fresh. Hard as they searched, not a single living stick was to be seen.

So there was only one thing left to do: repeat the climbing expedition at night—because the world's largest stick insect is known to be nocturnal. Equipped with headlamps and cameras, the climbers experienced something akin to a waking dream. Unbelievably, there in the middle of what was pretty much the sole bush on the entire stack sat twenty-four huge black stick insects staring at them.

Nobody knows how the insects made their way from Lord Howe Island to the sea stack some time before the extinction in 1920. If you can't fly or swim, a 12-mile trip across the open sea is quite a challenge. The most likely explanation is that eggs or a pregnant female hitched a ride with a bird or on floating vegetation and then the stick insects managed to survive for at least eighty years on the inhospitable sea stack, which is almost bare of vegetation.

We will draw a veil of silence over the bureaucracy that ensued. After two years of paper shuffling, permission was finally obtained for two males and two females to be fetched from the sea stack to form the start of a breeding program. Two of them (christened Adam and Eve, of course) survived, and healthy stocks of stick insects can now be found in several zoos, including in Europe.

But then the question arose of returning the rest of them to Lord Howe Island, where the species actually belonged—because a sea stack with a single bush at the mercy of rock falls is hardly fit to be a permanent home for a viable population of stick insects living in the wild. But on Lord Howe, the black rats still prevail. Unless they are wiped out, there's no point reintroducing the stick insects. And they aren't the only animals that would be glad to see the rats killed off: thirteen bird species and two reptile species face extinction unless the rats are eliminated. So the authorities now plan to do away with the rats once and for all. This will require extreme measures: 42 metric tons of poisoned cereal will be scattered over the island from a helicopter.

Of course, the process won't be entirely straightforward. First of all, animals other than the rats may die from eating the grain—including the birds people are trying to rescue. So the idea is to trap the most vulnerable bird species and keep them in a kind of temporary Noah's Ark, then release them again after the poison rain. But what consequences will this have for the genetic diversity of the birds, for example—because it won't be possible to capture all the individuals, will it?

And then some people are worried. There are only 350 human inhabitants on the island, but not all of them are keen on being showered with poisoned breakfast cereal, even though the authorities have assured them that no poison will be scattered near houses. Some probably also think that the big black stick insects are repulsive and no more deserving of protection than the black rats. Conservation biology is just as much about us humans and our thoughts and feelings as about the species we are trying to conserve.

New Times, New Species

In many ways nature is robust, and it is adapting all the time. New species emerge where we humans create new opportunities, such as deep below the ground in London, where the rough, damp tunnels of the Tube are home to a highly unusual mosquito. It belongs to the *Culex pipiens* species, the most common type of bloodsucking mosquito found in urban areas, but it has developed into a special genetic form (called *molestus*—"the troublesome one"), which is no longer capable of producing offspring with its mosquito relatives up in the light of day. What probably happened is that a couple of female mosquitoes found their way down into the depths at some point many years ago, perhaps when the Tube was being built in 1863. Since then, the London Underground mosquito has lived its own life down there, over thousands of mosquito generations.

The mosquito became notorious during the Second World War, when it was a source of great irritation to people seeking shelter in the Underground system during the Blitz. Today, hygienic standards are far better than in those days, and although roe deer, foxes, bats, woodpeckers, sparrow hawks, tortoises, and great crested newts have all been spotted in the tunnels, rats and mice are the main species that keep the few London Underground mosquitoes company.

Genetic analyses have shown that the mosquitoes' DNA varies among lines and stations: the Piccadilly Line mosquito is different from the Central Line mosquito—although not different enough to prevent the different Underground mosquito varieties from mating with one another. The leading theory is therefore that all of them are descendants of the same bold forebears from 150 years ago.

If it is true that the mosquito has developed into a new genetic form in just 150 years, it is an example of how evolution can work quickly now and then—as when populations are living in total isolation. Charles Darwin theorized that new species needed tens of thousands, if not hundreds of thousands, of generations to come into being. It is odd to think that while he sat pondering this in his house on the outskirts of London—he had just published *Origin of the Species* in 1859—a process of lighting-fast evolution might have been starting right beneath his feet.

We will probably see more such examples of new and rapid species formation in the future as a result of our intentional and unintentional displacement of species. The North American fly *Rhagoletis pomonella* used to live contentedly on hawthorn trees until apples came to the United States from Europe. Now the fly has two different genetic forms: one that eats only hawthorn berries and another that eats only apples. In just a couple of hundred years, one species is well on its way to becoming two. Even the parasite that lives on this fly is in the process of splitting into two species, one for the hawthorn-eating and one for the apple-eating larvae.

When new insects appear and others die out, the effect will depend on which species change—because, as I have shown in this book, different insects perform different tasks in nature. What's more, every insect is connected to other species through ingeniously adapted interactions, and this is the basis of all the goods and services nature offers us.

We humans have long taken the free services of insects for granted. Through intensive land use, climate change, insecticides, and the introduction of invasive species, we now risk altering conditions so quickly that insects will have difficulty delivering as they have done to date, despite nature's adaptability. On egotistical grounds alone, we should therefore be concerned about the health and well-being of these little creatures. Taking care of them is a form of life insurance for our children and grandchildren.

If we could just stop navel gazing for a second, we would see that this is about more than mere utility value. As far as we know, our planet is the only place in the universe where there is life. Many would say that we humans have a moral duty to rein in our dominance of the earth and give our millions of fellow creatures a chance to live out their tiny, wonderful lives, too.

Afterword

Somewhere far back in the mists of time we share a common forebear with the insects. Although the insects arrived long before us—they have a head start of several hundred million years—we also have a long shared history, for better and for worse. And there is no doubt that we need them. As the Harvard professor E. O. Wilson wrote, "The truth is that we need invertebrates but they don't need us. If human beings were to disappear tomorrow, the world would go on with little change. . . . But if invertebrates were to disappear, I doubt that the human species could live more than a few months."

This means we have everything to gain by caring a bit more about insects. I believe in knowledge, positive talk, and enthusiasm. Be curious about bugs, take the time to look and learn. Teach children about all the strange and useful things insects do. Talk nicely about bugs. Make your garden a better place for flower visitors. Let's get insects onto the agenda in land-use plans and official reports, agricultural regulations and state budgets. Let us take pleasure in colorful butterflies; let us admire the funny interactions among these little creatures and be thankful that insects step up to work on our behalf.

Insects are strange, intricate, funny, bizarre, fun, charming, and unique and never cease to amaze us. A Canadian insect researcher once said, "The world is rich in small wonders—but so poor in eyes that see them." My hope is that this book will open more people's eyes to the weird and wonderful world of insects—and the extraordinary lives they live alongside us on this planet we share.

Acknowledgments

I've had loads of great discussions about insects and related matters over the years. Thanks to my fantastic colleague Tone Birkemoe at NMBU for her unwavering enthusiasm, productive conversations, and comments on this text. And three cheers for all the other members of the NBMU's insect group, who all help talk up insects and make it a fun working environment. Thanks to my former colleagues at NINA (where I still have the pleasure of working part-time)—and to head of research Erik Framstad as a representative for everybody there—for all the stimulating conversations about absolutely everything between Heaven and Earth.

Thanks to my family, close and extended alike. My parents taught me to be curious about everything that moved outdoors in nature. I believe my mother has read, heard, watched, and said something nice about every single idea I've been involved with over the past few years. Thanks to my dear Kjetil for all the patience, tea, and buttered crispbread he has served up during late nights of writing. Thanks to our children, Simen, Tuva, and Karine, for all the fun we have together, and a special thank-you to Tuva for casting a sharp eye over the text and for wanting to illustrate the book and for drawing the chapter illustrations.It has been just incredible fun writing this book. To my Norwegian publisher, J.M. Stenersens Forlag, and editor, Solveig Øye, as well as my agent, Hans Petter Bakketeig at Stilton Literary Agency: Thanks for all the positive thinking and for taking such good care of me!

Last but not least, it has been incredible fun writing this book.

I've taken such joy in everything I've learned. And my publishers have been with me all the way. Thanks for that, and thanks for the support I received from the Norwegian Non-fiction Writers and Translators Association's Non-fiction Literary Fund.

Further Reading

These books have brought me great pleasure and inspiration, both generally and while writing this book. I recommend them to any readers keen to take a deeper dive into the fantastic world of the insects.

M. R. Berenbaum, *Bugs in the System* (Reading, MA: Addison-Wesley, 1995).

The books of Dave Goulson: *A Sting in the Tale* (2013), *A Buzz in the Meadow* (2014), and *Bee Quest* (2017), published by Picador and Jonathan Cape.

R. Jones, *Call of Nature: The Secret Life of Dung* (Exeter, UK: Pelagic Publishing, 2017.)

E. McAlister, *The Secret Life of Flies* (Buffalo, NY, 2017).

S. R. Shaw, *Planet of the Bugs: Evolution and the Rise of Insects* (Chicago: University of Chicago Press, 2014).

M. Zuk, *Sex on Six Legs: Lessons on Life, Love, and Language from the Insect World* (New York: Houghton Mifflin Harcourt, 2011).

Sources

INTRODUCTION

Andersen, T., V. Baranov, L. K. Hagenlund, et al. "Blind Flight? A New Troglobiotic Orthoclad (Diptera, Chironomidae) from the Lukina Jama–Trojama Cave in Croatia." *PLoS One* 11 (2016): e0152884.

Baust, J. G., and R. E. Lee Jr. "Multiple Stress Tolerance in an Antarctic Terrestrial Arthropod: *Belgica antarctica*." *Cryobiology* 24, no. 2 (1987): 140–47.

Berenbaum, M. R. *Bugs in the System*. Reading, MA: Addison-Wesley, 1995.

Bishopp, F. C. "Domestic Mosquitoes." US Department of Agriculture, leaflet no. 186, 1939.

Chapman, A. D. "Numbers of Living Species in Australia and the World" (2nd edition), Australian Biodiversity Information Services report for the Australian Biological Resources Study (2009).

Fang, J. "Ecology: A World without Mosquitoes." *Nature* 466 (2010): 432–34.

Huber, J. T., and J. S. Noyes. "A New Genus and Species of Fairyfly, *Tinkerbella nana* (Hymenoptera, Mymaridae), with Comments on Its Sister Genus *Kikiki*, and Discussion on Small Size Limits in Arthropods." *Journal of Hymenoptera Research* 32 (2013): 17–44.

"Hvor mange arter finnes i Norge?" From Hallvard Elven and Geir Søli, eds., *Kunnskapsstatus for Artsmangfoldet i Norge 2015* (Trondheim, Norway: Artsdatabanken, 2013). https://www.artsdatabanken.no/Pages/205713.

Kadavy, D. R., B. A. Plantz, C. A. Shaw, et al. "Microbiology of the Oil Fly, *Helaeomyia petrolei*." *Applied and Environmental Microbiology* 65, no. 4 (1999): 1477–82.

Kelley, J. L., J. T. Peyton, A.-S. Fiston-Lavier, et al. "Compact Genome of the Antarctic Midge Is Likely an Adaptation to an Extreme Environment." *Nature Communications* 5 (2014): 4611.

Knapp, F. W. "Arthropod Pests of Horses." In *Livestock Entomology*, ed. R. E. Williams, R. D. Hall, A. B. Broce, and P. J. Scholl. New York: Wiley, 1985, 297–322.

"Largest Species of Beetle." Guinness World Records. http://www.guinness worldrecords.com/world-records/largest-species-of-beetle/.

Leonardi, M. S., and R. L. Palma. "Review of the Systematics, Biology and Ecology of Lice from Pinnipeds and River Otters (Insecta: Phthiraptera: Anoplura: Echinophthiriidae)." *Zootaxa* 3630, no. 3 (2013): 445–66.

Misof, B., S. Liu, K. Meusemann, et al. "Phylogenomics Resolves the Timing and Pattern of Insect Evolution." *Science* 346 no. 6210 (2014): 763–67.

Nesbitt, S. J., P. M. Barrett, S. Werning, et al. "The Oldest Dinosaur? A Middle Triassic Dinosauriform from Tanzania." *Biology Letters* 9 (2013): 20120949.

Shaw, S. R. *Planet of the Bugs: Evolution and the Rise of Insects*. Chicago: University of Chicago Press, 2014.

"World's Longest Insect Discovered in China." Xinhuanet, May 5, 2016. http://news.xinhuanet.com/english/2016-05/05/c_135336786.htm.

Zuk, M. *Sex on Six Legs: Lessons on Life, Love, and Language from the Insect World*. New York: Houghton Mifflin Harcourt, 2011.

CHAPTER 1: SMALL CREATURES, SMART DESIGNS

Alem, S., C. J. Perry, X. Zhu, et al. "Associative Mechanisms Allow for Social Learning and Cultural Transmission of String Pulling in an Insect." *PLoS Biology* 14 (2016): e1002564.

Arikawa, K. "Hindsight of Butterflies." *BioScience* 51, no. 3 (2001): 219–25.

Arikawa, K., E. Eguchi, A. Yoshida, and K. Aoki. "Multiple Extraocular Photoreceptive Areas on Genitalia of Butterfly *Papilio xuthus*." *Nature* 288 (1980): 700–702.

Avarguès-Weber, A., G. Portelli, J. Benard, et al. "Configural Processing Enables Discrimination and Categorization of Face-Like Stimuli in Honeybees." *Journal of Experimental Biology* 213 (2010): 593–601.

Caro, T. M., and M. D. Hauser. "Is There Teaching in Nonhuman Animals?" *The Quarterly Review of Biology* 67, no. 2 (1992): 151–74.

Chapman, A. D. *Numbers of Living Species in Australia and the World*, 2nd ed. Canberra, Australia. Department of the Environment, Water, Heritage and the Arts.

Dacke, M., and M. V. Srinivasan. "Evidence for Counting in Insects." *Animal Cognition* 11 (2008): 683–89.

Darwin, C. "Charles Darwin's Beagle Diary," September 17, 1834. http://darwinbeagle.blogspot.no/2009/09/17th-september-1834.html.

Darwin, C. *The Descent of Man, and Selection in Relation to Sex*. London: John Murray, 1871.

Elven, H., and L. Aarvik. "Insekter Insecta." Naturhistorisk Museum, Universitetet i Oslo Artsdatabanken. https://artsdatabanken.no/Pages/135656.

"Eremitten flyttes til åpen soning." NINA (Norwegian Institute for Nature Research), August 25, 2017. http://www.nina.no/english/News/News-article/ArticleId/4321.

Falck, M. "La vevkjerringene veve videre." *Insekt-Nytt* 29 nos.1–2 (2004): 57–60.

Franks, N. R., and T. Richardson. "Teaching in Tandem-Running Ants." *Nature* 439 (2006): 153.

Frye, M. A. "Visual Attention: A Cell That Focuses on One Object at a Time." *Current Biology* 23 (2013): R61–63.

Gonzalez-Bellido, P. T., H. Peng, J. Yang, et al. "Eight Pairs of Descending Visual Neurons in the Dragonfly Give Wing Motor Centers Accurate Population Vector of Prey Direction." *Proceedings of the National Academy of Sciences* 110, no. 2 (2013): 696–701.

Gopfert, M. C., A. Surlykke, and L. T. Wasserthal. "Tympanal and Atympanal 'Mouth-Ears' in Hawkmoths (Sphingidae)." *Proceedings of the Royal Society B: Biological Sciences* 269, no. 1486 (2002): 89–95.

Jabr, F. "How Did Insect Metamorphosis Evolve?," *Scientific American*, August 10, 2012. https://www.scientificamerican.com/article/insect-metamorphosis-evolution/.

Leadbeater, E., and L. Chittka. "Social Learning in Insects—from Miniature Brains to Consensus Building," *Current Biology* 17, no. 16 (2007): R703–13.

Minnich, D. E. "The Chemical Sensitivity of the Legs of the Blow-Fly, *Calliphora vomitoria* Linn., to Various Sugars," *Zeitschrift für vergleichende Physiologie* 11, no. 1 (1929): 1–55.

Montealegre-Z, F., T. Jonsson, K. A. Robson-Brown, et al. "Convergent Evolution Between Insect and Mammalian Audition," *Science* 338, no. 6109 (2012): 968–71.

Munz, T. *The Dancing Bees: Karl von Frisch and the Discovery of the Honeybee Language*. Chicago: University of Chicago Press, 2016.

Ranius, T., and J. Hedin. "The Dispersal Rate of a Beetle, *Osmoderma eremita*, Living in Tree Hollows." *Oecologia* 126, no. 3 (2001): 363–70.

Shuker, K. P. N. *The Hidden Powers of Animals: Uncovering the Secrets of Nature*. London: Marshall Editions, 2001.

Tibbetts, E. A. "Visual Signals of Individual Identity in the Wasp *Polistes fuscatus*." *Proceedings of the Royal Society Series B: Biological Sciences* 269, no. 1499 (2002): 1423–28.

CHAPTER 2: SIX-LEGGED SEX

Banerjee, S., N. P. Coussens, F. X. Gallat, et al. "Structure of a Heterogeneous, Glycosylated, Lipid-Bound, In Vivo–Grown Protein Crystal at Atomic Resolution from the Viviparous Cockroach *Diploptera punctata*." *IUCrJ* 3, part 4 (2016): 282–93.

Birch, J., and S. Okasha. "Kin Selection and Its Critics." *BioScience* 65, no. 1 (2015): 22–32.

Boos, S., J. Meunier, S. Pichon, M. Kölliker, et al. "Maternal Care Provides Antifungal Protection to Eggs in the European Earwig." *Behavioral Ecology* 25, no. 1 (2014): 754–61.

Borror, D. J., C. A. Triplehorn, and N. F. Johnson. *An Introduction to the Study of Insects*. 6th ed. Philadelphia: Saunders College Publishing, 1989.

Brian, M. B. *Production Ecology of Ants and Termites*. Cambridge, UK: Cambridge University Press, 1978.

Eady, P. E., and D. V. Brown. "Male-Female Interactions Drive the (Un)repeatability of Copula Duration in an Insect." *Royal Society Open Science* 4 (2017): 160962.

Eberhard, W. G. "Copulatory Courtship and Cryptic Female Choice in Insects." *Biological Reviews* 66 (1991): 1–31.

Fedina, T. Y. "Cryptic Female Choice during Spermatophore Transfer in *Tribolium castaneum* (Coleoptera: Tenebrionidae)." *Journal of Insect Physiology* 53, no. 1 (2007): 93–98.

Fleming, N. "Which Life Form Dominates Earth?," BBC, February 10, 2015. http://www.bbc.com/earth/story/20150211-whats-the-most-dominant-life-form.
Hamill, J. "What A Buzz Kill: Male Bees' Testicles Explode When They Reach Orgasm." *The Sun*, October 6, 2016. https://www.thesun.co.uk/news/1926328/male-bees-testicles-explode-when-they-reach-orgasm/.

"Hjortelusflue." Folkehelseinstituttet, November 11, 2015. https://www.fhi.no/nettpub/skadedyrveilederen/fluer-og-mygg/hjortelusflue-/.

Lawrence, S. E. "Sexual Cannibalism in the Praying Mantid, *Mantis religiosa*: A Field Study," *Animal Behaviour* 43, no. 4 (1992): 569–83.

Lüpold, S., M. K. Manier, N. Puniamoorthy, et al. "How Sexual Selection Can Drive the Evolution of Costly Sperm Ornamentation." *Nature* 533, (2016): 535–38.

Maderspacher, F. "All the Queen's Men." *Current Biology* 17, no. 6 (2007): R191–95.

Nowak, M. A., C. E. Tarnita, and E. O. Wilson. "The Evolution of Eusociality," *Nature* 466, no. 7310 (2010): 1057–62.

Pitnick, S., G. S. Spicer, and T. A. Markow. "How Long Is a Giant Sperm?" *Nature* 375, no. 6527 (1995): 109.

Schwartz, S. K., W. E. Wagner, and E. A. Hebets. "Spontaneous Male Death and Monogyny in the Dark Fishing Spider." *Biology Letters* 9, no. 4 (2013): 20130113.

Shepard, M., V. Waddil, and W. Kloft. "Biology of the Predaceous Earwig *Labidura riparia* (Dermaptera: Labiduridae)." *Annals of the Entomological Society of America* 66 (1973): 837–41.

Sivinski, J. "Intrasexual Aggression in the Stick Insects *Diapheromera veliei* and *D. Covilleae* and Sexual Dimorphism in the Phasmatodea." *Psyche* 85, no. 4 (1978): 395–405.

Williford, A., B. Stay, and D. Bhattacharya. "Evolution of a Novel Function: Nutritive Milk in the Viviparous Cockroach, *Diploptera punctata*." *Evolution & Development* 6, no. 2 (2004): 67–77.

CHAPTER 3: EAT OR BE EATEN

Britten, K. H., T. D. Thatcher, and T. Caro. "Zebras and Biting Flies: Quantitative Analysis of Reflected Light from Zebra Coats in Their Natural Habitat." *PLoS One* 11, no. 5 (2016): e0154504.

Caro, T., A. Izzo, R. C. Reiner Jr., et al. "The Function of Zebra Stripes." *Nature Communications* 5 (2014): 3535.

Caro, T., and T. Stankowich. "Concordance on Zebra Stripes: A Comment on Larison et al. (2015)." *Royal Society Open Science* 2, no. 9 (2015) https://doi.org/10.1098/rsos.150323.

Darwin, C. "To Asa Gray 22 May 1860" (letter no. 2814). Darwin Correspondence Project. http://www.darwinproject.ac.uk/letter/DCP-LETT-2814.xml (1860).

Dheilly, N. M., F. Maure, M. Ravallec, et al. "Who Is the Puppet Master? Replication of a Parasitic Wasp-Associated Virus Correlates with Host Behaviour Manipulation." *Proceedings of the Royal Society B: Biological Sciences* 282, no. 1803 (2015). https://doi.org/10.1098/rspb.2014.2773.

Eberhard, W. G. "The Natural History and Behavior of the Bolas Spider *Mastophora dizzydeani* SP. n. (Araneidae)." *Psyche* 87, nos. 3–4 (1980): 143–69.

Haynes, K. F., C. Gemeno, K. V. Yeargan, et al. "Aggressive Chemical Mimicry of Moth Pheromones by a Bolas Spider: How Does this Specialist Predator Attract More than One Species of Prey?" *Chemoecology* 12, no. 2 (2002): 99–105.

Larison, B., R. J. Harrigan, H. A. Thomassen, et al. "How the Zebra Got Its Stripes: A Problem with Too Many Solutions." *Royal Society Open Science* 2, no. 1 (2015): 140452.

Libersat, F., and R. Gal. "What Can Parasitoid Wasps Teach Us About Decision-Making in Insects?" *Journal of Experimental Biology* 216, no. 1 (2013): 47–55.

Marshall, D. C., and K. B. R. Hill. "Versatile Aggressive Mimicry of Cicadas by an Australian Predatory Katydid." *PLoS One* 4, no. 1 (2009): e4185.

Melin, A. D., D. W. Kline, C. Hiramatsu, et al.. "Zebra Stripes through the Eyes of Their Predators, Zebras, and Humans." *PLoS One* 11, no. 3 (2016). e0145679.

Yeargan, K. V. "Biology of Bolas Spiders." *Annual Review of Entomology* 39 (1994): 81–99.

CHAPTER 4: INSECTS AND PLANTS

Babikova, Z., L. Gilbert, T. J. A. Bruce, et al. "Underground Signals Carried through Common Mycelial Networks Warn Neighbouring Plants of Aphid Attack." *Ecology Letters* 16, no. 7 (2013): 835–43.

Barbero, F., D. Patricelli, M. Witek, et al. "*Myrmica* Ants and Their Butterfly Parasites with Special Focus on the Acoustic Communication." *Psyche* 11 (2012): https://doi.org/10.1155/2012/725237.

Dangles, O., and J. Casas. "The Bee and the Turtle: A Fable from Yasuní National Park." *Frontiers in Ecology and the Environment* 10, no. 8 (2012): 446–47.

de la Rosa, C. L. "Additional Observations of Lachryphagous Butterflies and Bees." *Frontiers in Ecology and the Environment* 12, no. 4 (2014): 210.

Ekblom, R. "Smörbolls flugornas fantastiska värld." *Fauna och Flora* 102, no. 1 (2007): 20–22.

Evans, T. A., T. Z. Dawes, P. R. Ward, et al. "Ants and Termites Increase Crop Yield in a Dry Climate." *Nature Communications* 2 (2011) 2: 262.

Grinath, J. B., B. D. Inouye, and N. Underwood. "Bears Benefit Plants via a Cascade with Both Antagonistic and Mutualistic Interactions." *Ecology Letters* 18 (2015): 164–73.

Hansen, L. O. *Pollinerende insekter i Maridalen*. Årsskrift 2015. Grua: Maridalens Venner, 2015.

Hölldobler, B., and E. O. Wilson, *Journey to the Ants: A Story of Scientific Exploration*. Cambridge, MA: Belknap Press of Harvard University Press, 1994.

Lengyel, S., A. D. Gove, A. M. Latimer, et al. "Convergent Evolution of Seed Dispersal by Ants, and Phylogeny and Biogeography in Flowering Plants: A Global Survey." *Perspectives in Plant Ecology, Evolution and Systematics* 12 (2010): 43–55.

McAlister, E. *The Secret Life of Flies*. London: Natural History Museum, 2017.

Midgley, J. J., J. D. M. White, S. D. Johnson, et al. "Faecal Mimicry by Seeds Ensures Dispersal by Dung Beetles." *Nature Plants* 1 (2015): 15141.

Moffett, M. W. *Adventures among Ants. A Global Safari with a Cast of Trillions*. University of California Press, 2010.

Needham, J. *Science and Civilisation in China*. Vol. 6, *Biology and Biological Technology*. Part 1: *Botany*. Cambridge, UK: Cambridge University Press, 1986.

Oliver, T. H., A. Mashanova, S. R. Leather, et al. "Ant Semiochemicals Limit Apterous Aphid Dispersal." *Proceedings of the Royal Society B: Biological Sciences* 274, no. 1811 (2007): 3127–31.

Patricelli, D., F. Barbero, A. Occhipinti, et al. "Plant Defences Against Ants Provide a Pathway to Social Parasitism in Butterflies." *Proceedings of the Royal Society B: Biological Sciences* 282 (2015): 20151111.

"The Prickly Pear Story." Department of Agriculture and Fisheries. Queensland, Australia, 2016. https://www.daf.qld.gov.au/_data/assets/pdf_file/0014/55301/IPA-Prickly-Pear-Story-PP62.pdf.

Simard, S. W., D. A. Perry, M. D. Jones, et al. "Net Transfer of Carbon between Ectomycorrhizal Tree Species in the Field." *Nature* 388 (1997): 579–82.

Stiling, P., D. Moon, and D. Gordon. "Endangered Cactus Restoration: Mitigating the Non-target Effects of a Biological Control Agent (*Cactoblastis cactorum*) in Florida." *Restoration Ecology* 12, no. 4 (2004): 605–10.

Stockan, J. A., and E. J. H. Robinson, eds. *Wood Ant Ecology and Conservation: Ecology, Biodiversity and Conservation*. Cambridge, UK: Cambridge University Press, 2016.

Wardle, D. A., F. Hyodo, R. D. Bardgett, et al. "Long-Term Aboveground and Belowground Consequences of Red Wood Ant Exclusion in Boreal Forest." *Ecology* 92, no. 3 (2011): 645–56.

Warren, R. J., and I. Giladi. "Ant-Mediated Seed Dispersal: A Few Ant Species (Hymenoptera: Formicidae) Benefit Many Plants." *Myrmecological News* 20 (2014): 129–40.

Zimmermann, H. G., V. C. Moran, and J. H. Hoffmann. "The Renowned Cactus Moth, *Cactoblastis cactorum* (Lepidoptera: Pyralidae): Its Natural History and Threat to Native *Opuntia* Floras in Mexico and the United States of America." *Florida Entomologist* 84, no. 4 (2001): 543–51.

CHAPTER 5: BUSY FLIES, FLAVORSOME BUGS

"About Hornet Juice." Nature Sport Science Ltd. https://www.hornetjuice.com/what/.

Bartomeus, I., S. G. Potts, I. Steffan-Dewenter, et al. "Contribution of Insect Pollinators to Crop Yield and Quality Varies with Agricultural Intensification." *PeerJ* 2 (2014): e328.

Clegg, A. "Edible Insects: Grub Pioneers Aim to Make Bugs Palatable." *Financial Times*, February 17, 2015. https://www.ft.com/content/bc0e4526-ab8d-11e4-b05a-00144feab7de.

Crittenden, A. N. "The Importance of Honey Consumption in Human Evolution." *Food and Foodways* 19, no. 4 (2011): 257–73.

Davidson, L. "Don't Panic, but We Could Be Running Out of Chocolate." *The Telegraph*, November 17, 2014. http://www.telegraph.co.uk/finance/newsbysector/retailandconsumer/11236558/Dont-panic-but-we-could-be-running-out-of-chocolate.html.

DeLong, D. M. "Homopteran." Encyclopedia Britannica. https://www.britannica.com/animal/homopteran#ref134267.

Harpaz, I. "Early Entomology in the Middle East." In *History of Entomology*, ed. R. F. Smith, T. E. Mittler, and C. N. Smith. Palo Alto, CA: Annual Reviews, 1973.

Hogendoorn, K., F. Bartholomaeus, and M. A. Keller, "Chemical and Sensory Comparison of Tomatoes Pollinated by Bees and by a Pollination Wand." *Journal of Economic Entomology* 103, no. 4 (2010): 1286–92.

Isack, H. A., and H. U. Reyer. "Honeyguides and Honey Gatherers: Interspecific Communication in a Symbiotic Relationship." *Science* 243, no. 4896 (1989): 1343–46.

Klatt, B. K., A. Holzschuh, C. Westphal, et al. "Bee Pollination Improves Crop Quality, Shelf Life and Commercial Value." *Proceedings of the Royal Society B: Biological Sciences* 281, no. 1775 (2014): 20132440.

Klein, A.-M., I. Steffan-Dewenter, and T. Tscharntke. "Bee Pollination and Fruit Set of *Coffea arabica* and *C. canephora* (Rubiaceae)." *American Journal of Botany* 90, no. 1 (2003): 153–57.

Lomsadze, G. "Report: Georgia Unearths the World's Oldest Honey." Eurasianet, March 30, 2012. https://eurasianet.org/report-georgia-unearths-the-worlds-oldest-honey.

Ott, J. "The Delphic Bee: Bees and Toxic Honeys as Pointers to Psychoactive and Other Medicinal Plants." *Economic Botany* 52, no. 3 (1998): 260–66.

Spottiswoode, C. N., K. S. Begg, and C. M. Begg. "Reciprocal Signaling in Honeyguide-Human Mutualism." *Science* 353, no. 6297 (2016): 387–89.

"Språklig insekt i mat." Språkrådet, January 2015. http://www.sprakradet.no/Vi-og-vart/Publikasjoner/Spraaknytt/spraknytt-2015/spraknytt-12015/spraklig-insekt-i-mat/.

Totland, Ø., K. A., Hovstad, F. Ødegaard, and J. Åström. *Kunnskapsstatus for insektpollinering i Norge—betydningen av det komplekse samspillet mellom planter og insekter.* (Trondheim, Norway: Artsdatabanken, 2013).

Wotton, R. S. "What Was Manna?" *Opticon1826* 9 (2010). http://ojs.lib.ucl.ac.uk/index.php/up/article/download/1375/718.

CHAPTER 6: THE CIRCLE OF LIFE—AND DEATH

Barton, D. N., N. Vågnes Traaholt, S. Blumentrath, and R. Reinvang. "Naturen i Oslo er verdt milliarder. Verdsetting av urbane økosystemtjenester fra grønnstruktur." *NINA Rapport* 1113, February 2015. https://brage.bibsys.no/xmlui/bitstream/handle/11250/2371359/1113.pdf.

Cambefort, Y. "Le Scarabée dans l'Égypte ancienne. Origine et signification du symbole." *Revue de l'histoire des religions* 204 (1987): 3–46.

Dacke, M., E. Baird, M. Byrne, et al. "Dung Beetles Use the Milky Way for Orientation." *Current Biology* 23, no. 4 (2013): 298–300.

Direktoratet for Naturforvaltning. "Handlingsplan for utvalgt naturtype hule eiker." *DN Rapport* 1-2012, February 2012. http://www.miljodirektoratet.no/old/dirnat/attachment/2762/DN-rapport-1-2012_nett.pdf.

Eisner, T., and M. Eisner. "Defensive Use of a Fecal Thatch by a Beetle Larva (*Hemisphaerota cyanea*)." *Proceedings of the National Academy of Sciences* 97 (2000): 2632–36.

Evju, M. (red.), V. Bakkestuen, H. H. Blom, et al. "Oaser for artsmangfoldet—hotspot-habitater for rødlistearter." *NINA Temahefte* 61, June 2015. https://www.nina.no/archive/nina/PppBasePdf/temahefte/061.pdf.

Goff, M. L. *A Fly for the Prosecution: How Insect Evidence Helps Solve Crimes.* Cambridge, MA: Harvard University Press, 2001.

Gough, L. A., T. Birkemoe, and A. Sverdrup-Thygeson. "Reactive Forest Management Can Also Be Proactive for Wood-Living Beetles in Hollow Oak Trees." *Biological Conservation* 180 (2014): 75–83.

Jacobsen, R. M. "Saproxylic Insects Influence Community Assembly and Succession of Fungi in Dead Wood." PhD thesis, Norwegian University of Life Sciences, Ås, Norway, 2017.

Jacobsen, R. M., T. Birkemoe, and A. Sverdrup-Thygeson. "Priority Effects of Early Successional Insects Influence Late Successional Fungi in Dead Wood." *Ecology and Evolution* 5, no. 21 (2015): 4896–905.

Jones, R. *Call of Nature: The Secret Life of Dung*. Exeter, UK: Pelagic Publishing, 2017.

Ledford, H. "The Tell-Tale Grasshopper: Can Forensic Science Rely on the Evidence of Bugs?" *Nature*, June 19, 2007. http://www.nature.com/news/2007/070619/full/news070618-5.html.

McAlister, E. *The Secret Life of Flies*. London: Natural History Museum, 2017.

Parker, C. B. "Buggy: Entomology Prof Helps Unravel Murder." UC Davis, June 8, 2007. https://www.ucdavis.edu/news/buggy-entomology-prof-helps-unravel-murder/.

Pauli, J. N., J. E. Mendoza, S. A. Steffan, et al. "A Syndrome of Mutualism Reinforces the Lifestyle of a Sloth." *Proceedings of the Royal Society B: Biological Sciences* 281, no. 1778 (2014): 20133006.

Pilskog, H. E. "Effects of Climate, Historical Logging and Spatial Scales on Beetles in Hollow Oaks." PhD thesis, Norwegian University of Life Sciences, 2016.

Savage, A. M., B. Hackett, B. Guénard, et al. "Fine-Scale Heterogeneity Across Manhattan's Urban Habitat Mosaic Is Associated with Variation in Ant Composition and Richness." *Insect Conservation and Diversity* 8, no. 3 (2015): 216–28.

Storaunet, K. O., and J. Rolstad. "Mengde og utvikling av død ved i produktiv skog i Norge. Med basis i data fra Landsskogtakseringens 7. (1994–1998) og 10. takst (2010–13)." *Oppdragsrapport*. Ås: Norsk Institutt for Skog og Landskap, June 2015.

Strong, L. "Avermectins: A Review of Their Impact on Insects of Cattle Dung." *Bulletin of Entomological Research*, 82, no. 2 (1992): 265–74.

Suutari, M., M. Majaneva, D. P. Fewer, et al. "Molecular Evidence for a Diverse Green Algal Community Growing in the Hair of Sloths and a Specific Association with *Trichophilus welckeri* (Chlorophyta, Ulvophyceae)." *BMC Evolutionary Biology* 10 (2010): 86.

Sverdrup-Thygeson, A. (ed.), T. E. Brandrud (ed.), H. Bratli, et al. "Hotspots—naturtyper med mange truete arter. En gjennomgang av Rødlista for arter 2010 i forbindelse med ARKO-prosjektet." *NINA Rapport* 683, April 2011. https://www.nina.no/archive/nina/PppBasePdf/rapport/2011/683.pdf.

Sverdrup-Thygeson, A., O. Skarpaas, S. Blumentrath, et al. "Habitat Connectivity Affects Specialist Species Richness More than Generalists in Veteran Trees." *Forest Ecology and Management* 403 (2017): 96–102.

Sverdrup-Thygeson, A., O. Skarpaas, and F. Ødegaard. "Hollow Oaks and Beetle Conservation: The Significance of the Surroundings." *Biodiversity and Conservation* 19 (2010): 837–52.

Vencl, F. V., P. A. Trillo, and R. Geeta. "Functional Interactions Among Tortoise Beetle Larval Defenses Reveal Trait Suites and Escalation." *Behavioral Ecology and Sociobiology* 65, no. 2 (2011): 227–39.

Wall, R., and S. Beynon. "Area-Wide Impact of Macrocyclic Lactone Parasiticides in Cattle Dung." *Medical and Veterinary Entomology* 26, no. 1 (2012): 1–8.

Welz, A. "Bird-Killing Vet Drug Alarms European Conservationists." *The Guardian*, March 11, 2014. https://www.theguardian.com/environment/nature-up/2014/mar/11/bird-killing-vet-drug-alarms-european-conservationists.

Youngsteadt, E., R. C. Henderson, A. M. Savage, et al. "Habitat and Species Identity, Not Diversity, Predict the Extent of Refuse Consumption by Urban Arthropods." *Global Change Biology* 21 (2015): 1103–15.

Ødegaard, F., L. O. Hansen, and A. Sverdrup-Thygeson. "Dyremøkk—et hotspot-habitat. Sluttrapport under ARKO-prosjektets periode II." *NINA Rapport* 715, August 2011. https://www.nina.no/archive/nina/PppBasePdf/rapport/2011/715.pdf.

Ødegaard, F., A. Sverdrup-Thygeson, L. O. Hansen, et al. "Kartlegging av invertebrater i fem hotspot-habitattyper. Nye norske arter og rødlistearter 2004–2008." *NINA Rapport* 500, August 2009. https://www.nina.no/archive/nina/PppBasePdf/rapport/2009/500.pdf.

CHAPTER 7: FROM SILK TO SHELLAC

Andersson, M., Q. Jia, A. Abella, et al. "Biomimetic Spinning of Artificial Spider Silk from a Chimeric Minispidroin." *Nature Chemical Biology* 13, no. 3 (2017): 262–64.

Bower, C. F. "Mind Your Beeswax." *Catholic Answers*, November 1, 1991. https://www.catholic.com/magazine/print-edition/mind-your-beeswax.

Copeland, C. G., B. E. Bell, C. D. Christensen, and Lewis, R. V. "Development of a Process for the Spinning of Synthetic Spider Silk." *ACS Biomaterials Science & Engineering* 1, no. 7 (2015): 577–84.

Fagan, M. M. "The Uses of Insect Galls." *American Naturalist* 52 (1918): 155–76.

"Live Animals." FAOSTAT database, Food and Agriculture Organization of the United Nations. http://www.fao.org/faostat/en/#data/QA.

"Forskrift om endring i forskrift om tilsetningsstoffer til næringsmidler." Lovdata (Norwegian legal database), May 21, 2013. https://lovdata.no/dokument/LTI/forskrift/2013-05-21-510.

Koeppel, A., and C. Holland. "Progress and Trends in Artificial Silk Spinning: A Systematic Review." *ACS Biomaterials Science & Engineering* 3, no. 3 (2017), 226–37.

"De Nødvendige Tanninene." *Apéritif*, March 19, 2014. https://www.aperitif.no/artikler/de-nodvendige-tanninene/169203.

Oba, Y. "Insect Bioluminescence in the Post–Molecular Biology Era." In *Insect Molecular Biology and Ecology*. ed. Klaus H. Hoffman. Boca Raton, FL: CRC Press, 2015, 94–120.

Osawa, K., T. Sasaki, and V. Meyer-Rochow. "New Observations on the Biology of *Keroplatus nipponicus* Okada 1938 (Diptera; Mycetophiloidea; Keroplatidae), a Bioluminescent Fungivorous Insect." *Entomologie Heute* 26 (2014): 139–49.

Ottesen, P. S. "Om gallveps (Cynipidae) og jakten på det forsvunne blekk." *Insekt-Nytt* 25, nos. 1–2 (2000): 5–14.

Rutherford, A. "Synthetic Biology and the Rise of the 'Spider-Goats.'" *The Guardian*, January 14, 2012. https://www.theguardian.com/science/2012/jan/14/synthetic-biology-spider-goat-genetics.

Seneca the Elder, *Excerpta Controversiae* 2.7. Latin text and translation. http://perseus.uchicago.edu/perseus-cgi/citequery3.pl?dbname=Latin August2012&getid=0&query=Sen.%20Con.%20ex.%202.7.

Shah, T. H., M. Thomas, and R. Bhandari. "Lac Production, Constraints and Management—A Review." *International Journal of Current Research* 7, no. 3 (2015): 13652–59.

"Spinning Spider Silk Is Now Possible." Sveriges Lantbruksuniversitet, January 9, 2017. http://www.slu.se/en/ew-news/2017/1/spinning-spider-silk-is-now-possible/.

"Silk Industry: Statistics." International Sericultural Commission. from http://inserco.org/en/statistics.

Sutherland, T. D., J. H. Young, S. Weisman, et al. "Insect Silk: One Name, Many Materials." *Annual Review of Entomology* 55 (2010): 171–88.

"Tilsetningsforordningen: endrigsbestemmelser om bruk av stoffer på eggeskall." Europalov. http://europalov.no/rettsakt/tilsetningsforordningen-endringsbestemmelser-om-bruk-av-stoffer-pa-eggeskall/id-5444.

Tomasik, B. "Insect Suffering from Silk, Shellac, Carmine, and Other Insect Products." Essays on Reducing Suffering, January 29, 2017. http://reducing-suffering.org/insect-suffering-silk-shellac-carmine-insect-products/.

Wakeman, R. J. "The Origin and Many Uses of Shellac." https://www.antiquephono.org/the-origin-many-uses-of-shellac-by-r-j-wakeman/.

Zinsser & Co. *The Story of Shellac*. 14th printing. New York: Zinsser & Co., 1927.

CHAPTER 8: LIFESAVERS, PIONEERS, AND NOBEL PRIZE WINNERS

Aarnes, H. "Biomimikry." Store Norsk. Leksikon, 2016. https://snl.no/Biomimikry.

"About the Sleeping Chironomid." Sleeping Chironomid Research Group, 2009. http://www.naro.affrc.go.jp/archive/nias/anhydrobiosis/Sleeping%20Chironimid/e-about-yusurika.html.

Alnaimat, S. "A Contribution to the Study of Biocontrol Agents Apitherapy and Other Potential Alternative to Antibiotics." PhD thesis, University of Sheffield, 2011.

Amdam, G. V., and S. W. Omholt, "The Regulatory Anatomy of Honeybee Lifespan." *Journal of Theoretical Biology* 216, no. 2 (2002): 209–28.

Bai, L., Z. Xie, W. Wang, et al. "Bio-Inspired Vapor-Responsive Colloidal Photonic Crystal Patterns by Inkjet Printing." *ACS Nano* 8, no. 11 (2014): 11094–100.

Baker, N., F. Wolschin, and G. V. Amdam. "Age-Related Learning Deficits Can Be Reversible in Honeybees *Apis mellifera*." *Experimental Gerontology* 47, no. 10 (2012): 764–72.

"Biomimetic Architecture: Green Building in Zimbabwe Modeled After Termite Mounds." Inhabitat, November 29, 2012. http://inhabitat.com/building-modelled-on-termites-eastgate-centre-in-zimbabwe/.

Bombelli, P., C. J. Howe, and F. Bertocchini. "Polyethylene Bio-degradation by Caterpillars of the Wax Moth *Galleria mellonella*." *Current Biology* 27, no. 8 (2017): R292–93.

Carville, O. "The Great Tourism Squeeze: Small Town Tourist Destinations Buckle under Weight of New Zealand's Tourism Boom." *NZ Herald*, May 5, 2017. http://www.nzherald.co.nz/nz/news/article.cfm?c_id=1&objectid=11828398.

Chechetka, S. A., Y. Yu, M. Tange, et al. "Materially Engineered Artificial Pollinators." *Chem* 2, no. 2 (2017): 224–39.

Cornette, R., and T. Kikawada. "The Induction of Anhydrobiosis in the Sleeping Chironomid: Current Status of Our Knowledge." *IUBMB Life* 63, no. 6 (2011): 419–29.

Dirafzoon, A., A. Bozkurt, and E. Lobaton. "A Framework for Mapping with Biobotic Insect Networks: From Local to Global Maps." *Robotics and Autonomous Systems* 88 (2017): 79–96.

Drew, J., and J. Joseph. *The Story of the Fly: And How It Could Save the World.* Green Point, South Africa: Cheviot Publishing, 2012.

Dumanli, A. G., and T. Savin. "Recent Advances in the Biomimicry of Structural Colours." *Chemical Society Reviews* 45, no. 24 (2016): 6698–724.

"Eastgate Development, Harare, Zimbabwe." https://web.archive.org/web/20041114141220/http://www.arup.com/feature.cfm?pageid=292.

Fernández-Marín, H., J. K. Zimmerman, S. A. Rehner, and W. T. Wcislo. "Active Use of the Metapleural Glands by Ants in Controlling Fungal Infection." *Proceedings of the Royal Society B: Biological Sciences* 273, no. 1594 (2006): 1689–95.

Fly on the Wall: Making Fly Science Approachable for Everyone. http://blogs.brandeis.edu/flyonthewall/list-of-posts/.

Haeder, S., R. Wirth, H. Herz, and D. Spiteller. "Candicidin-Producing *Streptomyces* Support Leaf-Cutting Ants to Protect Their Fungus Garden Against the Pathogenic Fungus *Escovopsis*." *Proceedings of the National Academy of Sciences* 106, no. 12 (2009): 4742–46.

Hamedi, A., S. Farjadian, and M. R. Karami. "Immunomodulatory Properties of *Trehala* Manna Decoction and Its Isolated Carbohydrate Macromolecules. "*Journal of Ethnopharmacology* 162 (2015): 121–26.

Horikawa, D. D. "Survival of Tardigrades in Extreme Environments: A Model Animal for Astrobiology." In *Anoxia: Evidence for Eukaryote Survival and Paleontological Strategies*, A. V. Altenbach, J. M. Bernhard, and J. Seckbach. Dordrecht: Springer Netherlands, 2012, 205–17.

Hölldobler, B., and H. Engel-Siegel. "On the Metapleural Gland of Ants." *Psyche* 91, no. 3–4 (1984): 201–24.

"India Bank Termites Eat Piles of Cash." BBC News, April 26, 2011. http://www.bbc.com/news/world-south-asia-13194864.

"Infrared Sensor Systems and Devices." Google Patents. https://www.google.com/patents/US7547886.

King, H., S. Ocko, and L. Mahadevan. "Termite Mounds Harness Diurnal Temperature Oscillations for Ventilation." *Proceedings of the National Academy of Sciences* 112, no. 37 (2015): 11589–93.

Ko, H. J., C. H. Youn, S. H. Kim, and S. Y. Kim. "Effect of Pet Insects on the Psychological Health of Community-Dwelling Elderly People: A Single-Blinded, Randomized, Controlled Trial." *Gerontology* 62, no. 2 (2016): 200–209.

Kuo, F. E., and W. C. Sullivan. "Environment and Crime in the Inner City: Does Vegetation Reduce Crime?" *Environment and Behavior* 33, no. 3 (2001): 343–67.

Kuo, M. "How Might Contact with Nature Promote Human Health? Promising Mechanisms and a Possible Central Pathway." Frontiers in Psychology 6 (2015): 1093.

Liu, F., B. Q. Dong, X. H. Liu, et al. "Structural Color Change in Longhorn Beetles *Tmesisternus isabellae*." *Optics Express* 17, no. 18 (2009): 16183–91.

McAlister, E. *The Secret Life of Flies*. London: Natural History Museum, 2017.

North Carolina State University. "Tracking the Movement of Cyborg Cockroaches." EurekAlert!, February 27, 2017. https://www.eurekalert.org/pub_releases/2017-02/ncsu-ttm022717.php.

Novikova, N., O. Gusev, N. Polikarpov, et al. "Survival of Dormant Organisms after Long-Term Exposure to the Space Environment." *Acta Astronautica* 68, no. 9 (2011): 1574–80.

Pinar. "Entire Alphabet Found on the Wing Patterns of Butterflies." My ModernMet, November 13, 2013. http://mymodernmet.comkjell-bloch-sandved-butterfly-alphabet/.

Quackwriter. "A Breath of Maggoty Air." The Quack Doctor, October 15, 2016. http://thequackdoctor.com/index.php/a-breath-of-maggoty-air/.

Ramadhar, T. R., C. Beemelmanns, C. R. Currie, et al. "Bacterial Symbionts in Agricultural Systems Provide a Strategic Source for Antibiotic Discovery." *The Journal of Antibiotics* 67, no. 1 (2014): 53–58.

Sogame, Y., and T. Kikawada. "Current Findings on the Molecular Mechanisms Underlying Anhydrobiosis in *Polypedilum vanderplanki*." *Current Opinion in Insect Science* 19, no. 15 (2017): 16–21.

Sowards, L. A., H. Schmitz, D. W. Tomlin, et al. "Characterization of Beetle *Melanophila acuminata* (Coleoptera: Buprestidae) Infrared Pit Organs by High-Performance Liquid Chromatography/Mass Spectrometry, Scanning Electron Microscope, and Fourier Transform-Infrared Spectroscopy." *Annals of the Entomological Society of America* 94, no. 5 (2001): 686–94.

Van Arnam, E. B., A. C. Ruzzini, C. S. Sit, et al. "Selvamicin, an Atypical Antifungal Polyene from Two Alternative Genomic Contexts." *Proceedings of the National Academy of Sciences* 113, no. 46 (2016): 12940–45.

Wainwright, M., A. Laswd, and S. Alharbi. "When Maggot Fumes Cured Tuberculosis." *Microbiologist* (March 2007): 33–35.

Watanabe, M. "Anhydrobiosis in Invertebrates." *Applied Entomology and Zoology* 41, no. 1 (2006): 15–31.

Whitaker, I. S., C. Twine, M. J. Whitaker, et al. "Larval Therapy from Antiquity to the Present Day: Mechanisms of Action, Clinical Applications and Future Potential." *Postgraduate Medical Journal* 83, no. 980 (2007): 409–13.

Wilson, E. O. *Biophilia*. Cambridge, MA: Harvard University Press, 1984.

Wodsedalek, J. E. "Five Years of Starvation of Larvae." *Science* 46, no. 1189 (1917): 366–67.

World Economic Forum, Ellen MacArthur Foundation, and McKinsey & Company. *The New Plastics Economy: Rethinking the Future of Plastics*. 2016. https://www.ellenmacarthurfoundation.org/assets/downloads/EllenMacArthurFoundation_TheNewPlasticsEconomy_Pages.pdf.

Yang, Y., J. Yang, W. M. Wu, et al. "Biodegradation and Mineralization of Polystyrene by Plastic-Eating Mealworms: Part 1. Chemical and Physical Characterization and Isotopic Tests." *Environmental Science & Technology* 49, no. 20 (2015): 12080–86.

Yates, D. "The Science Suggests Access to Nature Is Essential to Human Health." University of Illinois, February 13, 2009. https://news.illinois.edu/blog/view/6367/206035.

Zhang, C.-X., X.-D. Tang, and J.-A. Cheng. "The Utilization and Industrialization of Insect Resources in China." *Entomological Research* 38 (2008): S38–47.

CHAPTER 9: INSECTS AND US AND AFTERWORD

Brandt, A., A. Gorenflo, R. Siede, et al. "The Neonicotinoids Thiacloprid, Imidacloprid, and Clothianidin Affect the Immunocompetence of Honeybees (*Apis mellifera* L.)." *Journal of Insect Physiology* 86, no. 1 (2016): 40–47.

Byrne, K. and R. A. Nichols. "*Culex pipiens* in London Underground Tunnels: Differentiation Between Surface and Subterranean Populations." *Heredity* 82, no. 1 (1999): 7–15.

Dirzo, R., H. S. Young, M. Galetti, et al. "Defaunation in the Anthropocene." *Science* 345, no. 6195 (2014): 401–6.

Dumbacher, J. P., A. Wako, S. R. Derrickson, et al. "Melyrid Beetles (*Choresine*): A Putative Source for the Batrachotoxin Alkaloids Found in Poison-Dart Frogs and Toxic Passerine Birds." *Proceedings of the National Academy of Sciences* 101, no. 43 (2004): 15857–60.

Follestad, A. "Effekter av kunstig nattbelysning på naturmangfoldet—en litteraturstudie." *NINA Rapport* 1081, 2014. http://www.nina.no/archive/nina/PppBasePdf/rapport/2014/1081.pdf.

Forbes, A. A., T. H. Q. Powell, L. L. Stelinski, et al. "Sequential Sympatric

Speciation across Trophic Levels." *Science* 323, no. 5915 (2009): 776–79.

Garibaldi, L. A., I. Steffan-Dewenter, R. Winfree, et al. "Wild Pollinators Enhance Fruit Set of Crops Regardless of Honeybee Abundance." *Science* 339, no. 6127 (2013): 1608–11.

Gough, L. A., A. Sverdrup-Thygeson, P. Milberg, et al. "Specialists in Ancient Trees Are More Affected by Climate than Generalists." *Ecology and Evolution* 5, no. 23 (2015): 5632–41.

Goulson, D. "Review: An Overview of the Environmental Risks Posed by Neonicotinoid Insecticides." *Journal of Applied Ecology* 50, no. 4 (2013): 977–87.

Hallmann, C. A., M. Sorg, E. Jongejans, et al. "More than 75 Per Cent Decline over 27 Years in Total Flying Insect Biomass in Protected Areas." *PLoS One* 12 (2017): e0185809.

IPBES. "Summary for Policymakers of the Assessment Report of the Intergovernmental Science-Policy Platform on Biodiversity and Ecosystem Services on Pollinators, Pollination and Food Production." Secretariat of the Inter-governmental Science-Policy Platform on Biodiversity and Ecosystem Services, Bonn, Germany, 2016.

McKinney, M. L. "High Rates of Extinction and Threat in Poorly Studied Taxa." *Conservation Biology* 13, no. 6 (1999): 1273–81.

Morales, C., J. Montalva, M. Arbetman, et al. "*Bombus dahlbomii*: The IUCN Red List of Threatened Species 2016" Red List, 2016. http://dx.doi.org/10.2305/IUCN.UK.2016-3.RLTS.T21215142A100240441.en.

Myers, C. W., J. W. Daly, and B. Malkin. "A Dangerously Toxic New Frog (*Phyllobates*) Used by Emberá Indians of Western Colombia, with Discussion of Blowgun Fabrication and Dart Poisoning." *Bulletin of the American Museum of Natural History* 161, article 2 (1978): 307–66.

Pawson, S. M., and M. K. F. Bader. "LED Lighting Increases the Ecological Impact of Light Pollution Irrespective of Color Temperature." *Ecological Applications* 24, no. 7 (2014): 1561–68.

Rader, R., I. Bartomeus, L. A. Garibaldi, et al. "Non-Bee Insects Are Important Contributors to Global Crop Pollination." *Proceedings of the National Academy of Sciences* 113, no. 1 (2016): 146–51.

Rasmont, P., M. Franzén, T. Lecocq, et al. "Climatic Risk and Distribution Atlas of European Bumblebees." *BioRisk* 10 (2015): 1–246.

Rogne, S., and S. B. Thoresen. "Vi kan nå genmodifisere mygg så vi kanskje kvitter oss med malaria for godt." *Aftenposten*, December 7, 2015. https://www.aftenposten.no/viten/i/4m90/Vi-kan-na-genmodifisere-mygg-sa-vi-kanskje-kvitter-oss-med-malaria-for-godt.

Sánchez-Bayo, F. & Wyckhuys, K. A. G. "Worldwide Decline of the Entomofauna: A Review of Its Drivers," *Biological Conservation* 232 (2019), pp. 8–27. https://doi.org/10.1016/j.biocon.2019.01.020.

Säterberg, T., S. Sellman, and B. Ebenman. "High Frequency of Functional Extinctions in Ecological Networks." *Nature* 499, no. 7459 (2013): 468–70.

Schwägerl, C. "Vanishing Act. What's Causing the Sharp Decline in Insects, and Why It Matters." *Yale Environment 360*, July 6, 2016. https://e360.yale.edu/features/insect_numbers_declining_why_it_matters.

Seibold, S. et al. "Arthropod Decline in Grasslands and Forests Is Associated with Landscape-level Drivers," *Nature* 574 (2019), pp. 671–4. 10.1038/s41586-019-1684-3.

Thoresen, S. B. "Gen-drivere—magisk medisin eller villfaren vitenskap?" Bioteknologierådet, June 1, 2016. http://www.bioteknologiradet.no/2016/06/gen-drivere-magisk-medisin-eller-villfaren-vitenskap/.

Tsvetkov, N., O. Samson-Robert, K. Sood, et al. "Chronic Exposure to Neonicotinoids Reduces Honey Bee Health Near Corn Crops." *Science* 356, no. 6345 (2017): 1395.

Vindstad, O. P. L., S. Schultze, J. U. Jepsen, et al. "Numerical Responses of Saproxylic Beetles to Rapid Increases in Dead Wood Availability Following Geometrid Moth Outbreaks in Sub-Arctic Mountain Birch Forest." *PLoS One* 9 (2014): e99624.

Vogel, G. "Where Have All the Insects Gone?" *Science*, May 10, 2017. http://www.sciencemag.org/news/2017/05/where-have-all-insects-gone.

Wilson, E. O. "The Little Things That Run the World (The Importance and Conservation of Invertebrates)." *Conservation Biology* 1, no. 4 (1987): 344–46.

Woodcock, B. A., J. M., Bullock, R. F. Shore, et al. "Country-Specific Effects of Neonicotinoid Pesticides on Honey Bees and Wild Bees." *Science* 356, no. 6345 (2017): 1393.

Zeuss, D., R., Brandl, M. Brändle, et al. "Global Warming Favours Light-Coloured Insects in Europe," *Nature Communications* 5 (2014): 3874.

Index

Index

Index

Index

Index